Environmental Management and Development

The continual degradation of the planet's environment is something that affects every country, be it developed or developing. Statements and policies are made at international levels, but their effectiveness is questionable. Important questions must be considered: Are techniques of environmental management used in the West appropriate and relevant to the needs and priorities of poorer developing countries? How can developing countries be expected to follow policies, or to have a responsible attitude to the environment, when the governments of developed countries have sidestepped policies and have repeatedly shrugged off any sense of global responsibility?

By focusing on environmental management, this book distinguishes itself from existing environment and development texts, whose emphasis is on listing problems, making warnings and voicing advocacy. The author moves on from these viewpoints to look at practical management and problem-solving techniques, including examining future challenges for our increasingly globalised society.

Environmental Management and Development clarifies the definition, nature and role of environmental management in development and in developing countries. It begins with an introduction to the key terms, issues and tools of environmental management, which are linked to and developed in the chapters on specific environmental issues, making extensive use of local case studies. The book concludes by discussing who pays for environmental management and its future in developing countries.

C. J. Barrow is a Senior Lecturer at University of Wales Swansea.

Routledge Perspectives on Development

Series Editor: Tony Binns, *University of Sussex*

The Perspectives on Development series provides an invaluable, up-to-date and refreshing approach to key development issues for academics and students working in the field of development, in disciplines such as anthropology, economics, geography, international relations, politics and sociology. The series will also be of particular interest to those working in interdisciplinary fields, such as area studies (African, Asian and Latin American Studies), development studies, rural and urban studies, travel and tourism.

Published:

Third World Cities, Second edition
David W. Drakakis-Smith

An Introduction to Sustainable Development, Second edition
Jennifer A. Elliott

Gender and Development
Janet Henshall Momsen

Rural–Urban Interactions in the Developing World
Kenneth Lynch

Environmental Management and Development
C. J. Barrow

Forthcoming:

Indigenous Knowledge and Development
Tony Binns, Peter Illgner and Étienne Nel

Health and Development
Hazel Barrett

Tourism and Development
Alison Lewis and Martin Elliott-White

Children and Development
Nicola Ansell

Water and Development
John Soussan and Mathew Chadwick

Theories of Development
Katie Willis

Environmental Management and Development

C. J. Barrow

Routledge
Taylor & Francis Group

LONDON AND NEW YORK

First published 2005
by Routledge
2 Park Square, Milton Park, Abingdon, Oxon OX14 4RN

Simultaneously published in the USA and Canada
by Routledge
270 Madison Avenue, New York, NY 10016

Routledge is an imprint of the Taylor & Francis Group

© 2005 C. J. Barrow

Typeset in Times by
HWA Text and Data Management, Tunbridge Wells
Printed and bound in Great Britain by
TJ International Ltd, Padstow, Cornwall

British Library Cataloguing in Publication Data
A catalogue record for this book is available from the British Library

Library of Congress Cataloging in Publication Data
A catalog record for this book has been requested

ISBN 0–415–28083–4 (hbk)
ISBN 0–415–28084–2 (pbk)

Contents

Figures

Boxes

Preface

Environmental management is evolving rapidly and is increasingly being applied in developing countries, and to transboundary and global issues. It is being embraced by many sectors of human activity and its practitioners are very diverse. Developed and developing countries share the same global environment and development activities are often linked; for example, a large car-hire company recently funded forest planting in Bangladesh in order to claim 'carbon-neutral' status on the world-scale. It was offsetting pollution it caused in the North against environmental 'improvement' in the South, which cost it less to undertake.

There have been few books on environmental management and developing countries, although there is no shortage of manuals on how to apply specific tools or approaches in particular situations.

This book explores environmental management, with the focus on developing countries and the development process worldwide. Environmental management is a wide field, and there is a great diversity of environmental managers involved in development. Expatriate specialist environmental consultants have a relatively tight focus and tend to use established tools and approaches. Academics in developed and developing countries may research issues in-depth, but usually deal less with practical decision making. Commercial or government environmental managers have their priorities. Stakeholders in the development process include individuals, groups, nations, special interest groups, and many more.

The process of development takes place in the environment, using resources, generating waste and causing other impacts. Efforts are now made to plan and control the development process and how it relates with nature, but what exactly is environmental management? Is it a single field or discipline? Is it a process? Is it an agreed approach? Is it efforts to identify and pursue goals? Perhaps a philosophy? Or, is it environment and development problem solving?

Humans have survived environmental challenges and prospered as a species through forward thinking and being adaptable – environmental management is, or should be, the latest means to those ends. In a crowded and increasingly degraded, and possibly unstable world – environmentally, socially and economically – reacting to challenges after they materialise could be too late; moving to safe localities if a problem develops is no more an option with crowding and global threats; people today are probably less adaptable than past generations, and large populations also mean more time is needed to make many adaptations. Environmental management supports proactive planning, forward thinking and co-ordination of diverse activities – it is going to be a key element in future human fortunes.

Chris Barrow
University of Wales Swansea
May 2004

Part I
Theory and approaches

1 Introduction

Key chapter points

- This chapter explores the meaning of 'development' and 'environmental management' and how they interrelate.
- The evolution of interest in development and environmental issues and their management are reviewed.
- Environmental management helps steer the transition from environmental exploitation and largely retrospective responses to challenges and possibilities, to stewardship of nature and proactive assessment of threats and opportunities.

In the 1960s the media began to show satellite photographs of the world. This more or less coincided with several large environmental accidents, Cold War fears of nuclear annihilation, recognition by some that frontiers were closing and unused land was limited, and the birth of an environmental movement which was voicing concern about overpopulation, pollution and other development ills. The catchphrase 'spaceship Earth' was coined – implying a vulnerable, unique and shared life-support system, with a wayward set of passengers who tend to forget that what one does affects all. On any ship the hope is that it can be safely steered, that the crew know what they are doing, act responsibly, and manage to control the passengers. Like a 1920s ocean liner, 'spaceship Earth' has different classes of passengers: relatively few first-class, living in luxury – mainly those of the rich nations but also some in poor countries; more numerous second-class assigned to rather harsh conditions – mainly citizens of the developing countries; and 'stowage-class' – the unemployed or underemployed of urban slums, landless rural poor and those marginalised in remote and harsh environments. The task of communicating with all these passengers and managing 'spaceship Earth' falls to environmental managers. Like seamanship it is as much an art as a science, involves a diversity of specialists, demands many skills, and at times must deal with wholly unexpected and unfamiliar problems.

Development

Environmental management and development are both difficult to define. The former can be a goal or vision, an attempt to steer a process, the application of a set of tools, a more philosophical exercise seeking to identify and establish new outlooks, and much more. Individual environmental managers may have a problem-solving, sectoral, local, regional or global focus. They may be academics, regional or national decision makers and planners, non-governmental organisation (NGO) staff, company executives, international civil servants, or all sorts of individuals or groups who are environmental stakeholders in some way using natural resources – herders, farmers, fishermen and so on. Development can be a goal or a vision, application of a set of tools, or an attempt to steer a process. Developers are as equally varied a set of groups or individuals as environmental managers. Often the two fields overlap, and in both cases the outlook of the individual very much influences goals and approaches. Both environmental management and development are fields which demand a multidisciplinary view and make it possible for different disciplines, religions, classes, ethnic groups, political outlooks and genders to come together and seek mutually beneficial approaches to important issues. Environmental management and development often move from advocacy to actually trying to achieve goals; and both review current and possible future scenarios to try and identify routes toward better conditions. In the last 40 years or so, both environmental management and development have had to address global issues; sustainable development has become a goal for both. Before the 1960s both environmental management and development were top–down activities, but they now seek popular participation or even empowerment: the environment and people were once to be conquered, today they are to be understood and worked with.

The world today has a rich North and a poor South with pockets of poverty in the former and some very affluent people in the latter. Most of the world's population aspires to the material lifestyles and consumption patterns apparent in richer nations. Others may be less secular and look forward to non-material 'development' in the shape of an increase in contentment, more sense of security, religious or cultural enrichment, or whatever. The former, material, outlook is dominant and with growing populations the question is, will the Earth's environment support these hopes? The environmental manager is concerned with exploring what can be done to improve people's lifestyles, given the structure and function of the environment, and then implementing it. Some countries have achieved what they and others see as development, thanks to one or more of the following: agriculture, science, industrial expansion, natural resource exploitation, colonial expansion, trade and intellectual skills. Development is thus widely seen as a goal and an ongoing process, but there is less certainty over its exact meaning or how it functions, or the strategy that is best adopted to pursue it.

Providing a universally acceptable definition of development is impossible, although most would accept that it is a process of change which can accelerate, slow, stagnate or run backwards. Planners, managers, intellectuals and inspired individuals can try to drive development forward in a wide range of ways, such as planning, key speeches, books, fashions, inspirational acts and many other interventions. Definitions reflect the current values of those making them (Adams, 2001: 1; Sachs, 1992: 2). Most adopt an anthropocentric viewpoint, placing humans first, so that development means environment takes second place; however, there are environmentalists who are more ecocentric, i.e. regard environmental care to be at least as important as human needs. The establishment, including scientists, has also started to treat the Gaia Hypothesis more seriously – which means human needs must be weighed against maintaining critical natural processes. Thus, what was deemed development in the past may not be seen to be so today – and, at any given moment in time, individuals, groups and societies differ in their conception of what it is.

Some see development as more a learning or evolutionary process, often with the West assumed to be the currently most advanced stage (Dupâquier and Grebenik, 1983: 247). Geographers and geomorphologists have tried to understand and model landscape development; biologists explore phylogeny and ontogeny; ecologists explore ecosystem development; psychologists and educationalists focus on personality development and learning. Most of these fields have been influenced by the concept of evolution proposed by Darwin in the 1860s, and view development as a process with stages, rules and predictability. A typical Western definition used by those working with developing countries would be: ... *an ambiguous term for a multidimensional process involving material, social and organisational change, accelerated economic growth, the reduction of absolute poverty and inequality.* Chambers (1997) was more succinct; suggesting it was 'good change'.

Development, it has been suggested, is the economic component of a wider process: modernisation (Brookfield, 1973; Mabogunje, 1980). Modernisation today is generally accepted to be a change toward economic, social and political systems developed in Western Europe and North America between the seventeenth and nineteenth centuries, and, since the 1960s, increasingly around the Pacific Rim. In theory, a group of people could develop in a range of ways, but remain unmodernised, or vice versa. Athanasiou (1997: 19) suggested development was: '... what remains after the vapours of progress have boiled away'. Whatever the precise definition, development and modernisation – at the national, regional or local level – may be driven from within, but are linked in varying degree to external forces. Environmental management has, according to a well-used cliché, to 'act locally and think globally' – international agreements and transboundary problems (they cross borders) have to be considered.

Efforts to improve human material well being and security have often been poorly managed, and civilisations have seldom lasted for more than a thousand years before human or environmental problems or both have confounded them. Generally, the group in power has decided fashions and desirable goals (these have not always been material things) and they are often out of touch with nature and the rest of society. The idea that humans could and should shape their world to improve well being before death was little voiced before the sixteenth century AD. This was probably because lives were comparatively short and the ability to challenge the environment was limited, so most accepted hardship and disasters as the will of the gods or God; also, religious authorities and states typically frowned on challenges to the status quo. In the West an English statesman and theologian, Thomas More, was one of the first to publish views on how humans might develop in the early sixteenth century, in his book *Utopia*.

Various religions until recently saw the world as created for humans to use and benefit from. Unfortunately, in the West this tended to prompt exploitation rather than stewardship – greed and a desire for economic growth have driven development, and it has been likened to 'the ideology of a cancer cell'. One could build on this and regard colonial expansion as the worldwide metastasis, a catastrophic spread to form secondary cancers. From roughly 1650 in Europe the belief took hold that material work could improve humanity, rather than only through religious works; and from the 1750s scientific enquiry and rationalism were harnessed (Uglow, 2002). The results were impressive, and today a Eurocentric, liberal democratic, rational and scientific bias, which is broadly true of environmental concern since the 1960s, runs throughout much of the world's development activity. The currently dominant Western outlook is mainly anthropocentric, placing human needs (and often profit) before protection of the environment.

There is a hope that humans can control their development, stretch Nature to optimise resource use, and avoid environmental disaster – some would see that as the goal of environmental management. Development is increasingly seen as requiring reduction of intergroup disparity or a 'social transformation' through the use of capital, technology and knowledge. Clearly, it is difficult to separate concepts of development from cultural, ethical and religious outlook. Before the twentieth century, Westerners would not have referred to 'development'; they talked of conquering Nature, 'civilising' (moral improvement), and progress. By the mid-nineteenth century Westerners were spreading their laws, encouraging Christian missionary work, and selectively supporting education, politics and governance, the development of communications, technology and especially trade. The expression 'development' only appeared in the mid- to late-1940s, at first mainly measured by economic criteria like per capita growth, income and degree of industrialisation. In 1949 United States (US) President Harry Truman referred to it as intervention by 'liberal democracies' in the southern hemisphere's 'under-developed areas' which had replaced 'imperial exploitation', and probably as part of the containment of communism (Said, 1994). In practice, the USA had been experimenting with development – the State intervening to 'steer', rather than relying on *laissez-faire* progress – since the 1930s or earlier. For example, between 1933 and 1936 the Tennessee Valley Authority included resource management, environmental care and social improvement efforts, and in pursuit of these orchestrated state, federal and private resources. The Marshal Plan after the Second World War can also be seen as a thought-out development programme. In Europe and its colonies there were also efforts at planned development, and some environmental awareness. However, everywhere development was something that was mainly done by economists, either directly by promoting economic growth, or indirectly by commissioning and steering technology, agricultural development and mining. Better management of the environment had a low priority because it promised limited profit and Nature was still generally seen to be unlimited and resilient.

Escobar (1992) argued that development is a mechanism used by richer nations for the management of the poor since the mid-1940s. Starting in the 1950s, but much more after the 1970s, it has been seen to involve social improvement, and is increasingly judged by non-economic benchmarks: literacy rates, child survival, average age at death and suchlike. Alongside this there has been growing concern for environmental quality. Fifty years ago development was measured with economic indices; since the early 1990s multi-factor indices like the Human Development Index or the Amnesty International 'Freedom Index' or those of the World Development Institute have been used (UNDP, 1991). Growing concern for environmental quality and interest in pursuing sustainable development has prompted the inclusion of sustainability or environmental indices into development measurement.

Accurate measurement of development is important. Although his writing has been greeted with considerable criticism, Lomborg (2001) makes a valuable point: that too many people make selective and mistaken or misleading use of environmental and developmental evidence. Environmental managers must ensure discussion; negotiations and policy making are not based on misconceptions or false statistics ('myths'). They may have to work with inadequate knowledge and data.

Should richer countries; international agencies, NGOs and charismatic leaders 'assist' others to develop and manage their environments? It could be argued that people must do it for themselves, and there are countries that have considered or embarked on 'decoupling' their development from the rest of the world (Adams, 1990: 72, 83). However, problems in one country affect others, and it may require aid, at least at first, to resolve them. Environmental problems are often caused by poverty and can have global impacts. After 1945, the Eastern and Western blocs promoted their different styles of development; the Third World got caught up in the rivalry, or was neglected to pay for it. Both Western and Eastern blocs have generated serious environmental problems. Following the end of the Cold War in the late 1980s there is hopefully more funding to fight poverty and improve environmental management, and fewer restrictions on gathering and exchanging data, monitoring and policing the environment.

Between the 1870s and 1930s some supporters of social Darwinism and environmental determinism (see Chapter 2) saw underdevelopment to be rooted in racial or environmental endowment (Bernstein, 1973: 13–30; Hettne, 1990; Corbridge, 1991; Barrow, 2003: 17–31). More recently, failure of countries to develop has been attributed to a range of factors including: dependency in relation to trade, aid and technology; colonisation; decolonisation; neo-colonialism; or the side effects of the Cold War (Toye, 1987: 1–21). There is currently a 'post-development' school of thought, which claims that efforts to develop have failed and so the concept has been discredited.

From 1949 to the early 1970s development was concerned primarily with the reduction of poverty; environmental concern was deemed irrelevant, a 'luxury' poor people could not afford, or was even seen as part of a conspiracy to hold back the less-developed nations. It was not until around 1987 that it was widely accepted that development needed effective environmental management and vice versa.

Since the 1970s the view which has increasingly attracted attention is that humankind has a limited time to set in motion development that will sustain the world's people indefinitely with a satisfactory quality of life (Caldwell, 1977: 98; Berger, 1987: 116; Ghai and Vivian, 1992). En route to that goal it will probably be necessary to support too large a population and to cope with associated environmental damage and conflicts, perhaps for several decades. This 'overshoot' viewpoint was partly prompted by the publication in 1972 of the Club of Rome's *The Limits to Growth* (Meadows et al., 1972), which warned that human demands could outstrip global limits with catastrophic consequences unless there was effective environmental and development management. Initially the reaction was to urge 'zero growth', essentially the stopping of further development, which was greeted with dismay. Modelling suggested appropriate environmental and development management could avert disaster and allow some livelihood improvements – around this time the concept of sustainable development appeared and gained support. This was a way of somehow allowing development without exceeding the limits; as an alternative to zero growth or disaster it was most welcome. The concept was popularised by the Brundtland Report (World Commission on Environment and Development, 1987) and there is little sign of any diminishment of support for it.

At the risk of airing another cliché, roughly three-quarters of the world's population (ca 6,200 million) subsist on about one-quarter of the total income, and mainly live in 141 developing countries – the South, or the Third World. Developing countries are a very diverse group, in terms of degree of poverty, natural resource endowment, environment, politics, culture and in many other respects. A common, but not universal, characteristic is that they are sub-tropical or tropical. The 28 developed countries – the North, or the First World – include seven major industrialised nations: Canada, France, Germany, Italy, Japan, the United Kingdom (UK), and the USA (the 'G7'countries), and a number of states are close to, or heading toward, developed country status. So far development has been largely a Western phenomenon, but with industrialisation and technical capabilities growing in India and China, there is a strong chance countries of the Pacific Rim will dominate in the future; and environmental problems will emerge there.

Development management

Development management has evolved independently of environmental management, but often overlaps. Development management is essentially the manipulation of interventions aimed at promoting development (Thomas, 1996; Kirkpatrick et al., 2002). Development managers focus on interorganisational relationships, seeking 'partnerships' and 'co-operation', and are skilled in dealing with political agendas and in co-ordination. Environmental managers seldom combine ecological and development management skills sufficiently, so they often need to work closely with development managers. For much of the history there has been strong support for *laissez faire* approaches, rather than interventions. Fashions and needs change – *laissez faire* development is not wise in a crowded and vulnerable world – development management and environmental management are vital.

Current ideas about development and environmental management are much influenced by Western liberal democracies, especially the USA. In some developing countries the established legal system, civil engineering regulations and methods of governance are heavily derived from the West, and some of their environmental problems are a consequence of this. For example, water laws evolved in wet temperate Europe and transferred to developing countries are often unsuitable for drier environments and different socio-economic conditions. Environmental managers must remodel approaches, ethics, laws and so on, to better suit developing countries. This may take time because experts are still often trained in developed countries.

Recent development has taken place during relatively stable and benign environmental conditions and this may not last. There is a rapidly increasing human population which stresses the environment. Estimates place world human population of about 1,000 years ago at between 20 and 26 million; by 1500 AD it had probably risen to between 400 and 500 million; now it has exceeded 6,200 million (McNeill, 2000: 7). There are 'developmental' stresses on the environment: since the 1980s structural adjustment programmes, rising oil prices and debt have reduced the funds available to many countries to deal with pollution, conservation and other challenges. Changes like loss of social capital, globalisation, capital penetration and technological innovation have also been causing environmental and human welfare problems (Smil, 1987; Redclift, 1995). The challenges faced by development management and environmental management are growing.

Growing interest in environmental issues

Little, if any, of the world, is nowadays wholly 'natural', in the sense that it is unaltered by human activity. An eminent scientist recently suggested that the current geological unit, the Holocene, should be ended and succeeded by the Anthropocene, or 'human-altered', Period. Humans are not the only organisms to alter the environment; photosynthetic algae and bacteria formed the Earth's free atmospheric oxygen roughly 2,500 million years ago, and have maintained it at about 21 per cent of the total gas mix ever since. Also, marine plankton can affect cloud cover, and remove carbon from the atmosphere and deposit it as carbonates on the seabed. Life and the abiotic environment are in all probability unconsciously working together maintaining balances (the Gaia Hypothesis). However, humans uniquely have the potential to consciously interact with the environment and to pro-actively manage it. Whether humans realise their potential or continue to neglect, mismanage and destroy the global balance remains to be seen. Environmental management seeks to achieve that potential to maintain a global balance, and if possible, improve people's well being. It has been argued that a crisis is close, and that there is limited time available to get environmental management right before the disrupted Gaian system shifts to a state disastrously unfavourable to humans. Various estimates suggest there is no more than a generation or two available – The Brundtland Report (World Commission on Environment and Development, 1987:

8) observed that: 'Most of today's decision-makers will be dead before the planet feels the heavier effects of acid precipitation, global warming, ozone depletion. Most of today's young voters will still be alive.'

Those warning of a crisis cite a number of causes, including: population growth, greed, consumerism, 'faulty' Western development ethics, ignorance, and careless use of technology (Ehrlich and Ehrlich, 1990; Barkey, 2000). Blame is laid at the feet of both developed and developing countries, the relative share depending on who is making the judgement. The citizenry of developed countries may be more numerous and increasing more rapidly, but developed country citizens consume much more per capita and produce greater quantities of waste: for example, an average American had roughly 280 times the environmental impact of one Rwandan in 2003.

There are various reactions to environmental degradation:

1   Ignore the threat;
2   Advocate abandonment of technology and a return to simple ways;
3   Use all 'tools' available, including technology, education and establishment of new ethics, to achieve sustainable development.

The first two will result in disaster – there is no way that over 6,200 million people can return to a pre-technical 'Eden' and survive. Careful use of technology must be part of the way forward (Lewis, 1992). Humans have caused so much damage that rehabilitation will demand their efforts, rather than reliance on Nature. I feel there is no choice other than the third, and the task of environmental managers is to steer this.

The idea that the world faces an environmental crisis may provoke valuable changes, but it can also encourage emotive, journalistic debate, 'fire-fighting' solutions, ill-considered short-term focus approaches, and activities that divert attention from other important tasks. If causes and treatments are not carefully researched, little will be achieved. Roe (1995) explored why 'crisis' is frequently recognised in Africa; where land – according to foreign experts – is always either 'overpopulated' or 'underpopulated'. 'Crisis' is something of an ongoing 1970s cultural construct (Firkis, 1993); it has served to focus public attention on important issues, but it may well be damaging when workable strategies are needed. Those warning of crisis have frequently been branded pessimistic 'Cassandras'. The majority of people in rich and poor countries don't think much about threats; a few optimistic 'cornucopians', like the biologist E.O. Wilson, believe humanity will successfully rise to any challenge and survive (Wilson, 2002). Environmental management must watch for threats, *accurately* assess which are genuine, decide priorities, and then strive to get appropriate action to avert or better survive problems.

Pollution, soil erosion, overfishing and excess hunting, loss of forests and other problems are presently apparent and, looking back, environmental historians, palaeoecologists, archaeologists and geologists expose plenty of past environmental disasters. For half a century attention has focused on how humans affect the environment – especially global warming – rather than how it affects humans: this is unwise. Awareness of the past helps environmental managers assess future scenarios; the information can also boost public interest in environmental forecasting and pro-active management.

Between the eighteenth century and the late 1940s the prevailing viewpoint in the West was that Nature was to be studied, catalogued, tamed and exploited, and that the Earth was virtually limitless and resilient (even after the USA Midwest Dust Bowl disaster of the 1930s). The frontier was still 'open', with land to settle and few signs of environmental stress, other than localised pollution and some loss of biodiversity. The outlook, still not fully altered, was essentially mechanistic – that Nature was relatively easy to understand, model and control – and there was little awareness of the complexity, vulnerability and limitations of the Earth's ecosystems. In the 1960s attitudes began to change.

During the last four decades there has been a marked development of interest in environmental issues by those outside of the 'academic' community. This has been prompted by a complex of causes which include: increasingly apparent pollution; loss of biodiversity; declining fish stocks; soil degradation; deforestation; a realisation that the world is finite and easily damaged; concern at the rate of human population growth; and worries about the threat of nuclear warfare and inadvertent technological disaster.

Since roughly in the 1960s interest in environmental issues has increased; some refer to this as an 'environmental movement', but the label 'environmentalism' is widely used for it. The followers of environmentalism – environmentalists – are a very diverse group who espouse a wide variety of values, but all regard ecological concern as crucial. Environmentalism largely developed in Western 'liberal democracies', notably west-coast USA and Western Europe; but has been embraced by growing numbers in the rest of the world (see Box 1.1). From the late 1980s there has been worldwide expansion of literature and media coverage of environment and development issues, and the growth of educational courses and agencies concerned with the field (Adams, 1990; Gupta, 1988; Gupta and Asher, 1998).

Sustainable development had become a prominent part of environment and development discourses by the twenty-first century, and various authorities have noted that it is a concept that helps integrate environmental management and development management. Caution is needed; sustainability and sustainable development are not the same, but are often used as if they were. The former is the ongoing function of an ecosystem or use of a resource, and implies steady demands; the latter is improvement of well-being and lifestyles and, in the foreseeable future, implies growing demands.

Sustainable development is a goal, but people are likely to resist changing environmentally damaging lifestyles if it means paying more for necessities or even luxury items; and many, through poverty, are unable to do so. There are governments and businesses that have genuinely embraced environmental care; however, some are ineffective and achieve little; there are businesses that hijack it for their own ends, and authorities that ignore it for economic or 'strategic' reasons. Social changes are needed, and economics, governance and law have to evolve to support environmental management. Optimistic forecasts assume humans will progress toward less damaging ways; others are more pessimistic and, perhaps

Box 1.1

An outline of the evolution of post-1960s environmental concern

Mid 1960s to mid 1970s

Between the 1960s and mid 1970s the focus was largely on pollution (in some part due to Rachel Carson's book *Silent Spring*, 1962), the population 'explosion' and the careless use of technology. One of the more influential publications of the period was *The Limits to Growth* (Meadows et al., 1972), which voiced Malthusian concerns that development was exceeding environmental limits and that there would be disaster if adequate corrective action were not taken in time. Most of the environmentalist activity consisted of warnings, efforts to prompt debate and advocacy. On the whole, environmental concern was not politicised before the 1980s, and in North America still relatively little after that.

The 1972 United Nationas (UN) Conference on Human Development (Stockholm) was the first major international meeting to address environment and development issues. The United Nations Environment Programme (UNEP), the first international UN environmental agency, was established in 1973. However, there was widespread suspicion of environmental concern, and it was common for it to be seen as hindering development or as part of a conspiracy to hold back poor nations. Even more widespread was the attitude that environmental concern was a 'luxury' which developing countries were too poor to address. Few countries had environmental ministries and there was limited media and public interest in rich or poor nations.

Mid 1970s to 1992

Many countries established environmental ministries, and agencies established environmental departments; business became interested, the media acquired green-issues editors, environmentalist NGOs and international bodies sprang up and became more influential. By this time, at least in the richer nations, the public was becoming more familiar with environmental issues. There was a broadening application, popularisation and politicisation of environmentalism – 'greening' – which Adams (2001: xvii) argued seeks to bridge the gulf between environmentalism and development and between theory and practice. One of the more influential publications of this period was the Brundtland Report (World Commission on Environment and Development, 1987), which helped propagate and establish the concept of sustainable development (Royal Tropical Institute, 1990; Erocal, 1991).

Although interest had been aroused by the 1970s it really took until the late 1980s before there was significant developing country involvement in environmental bodies like the World Conservation Union (IUCN), UNEP, the World Wide Fund for Nature or Greenpeace (Holdgate, 1999: vii). Citizen group action in the USA in the late 1970s forced the main American aid agency USAID to apply environmental assessment before financing developing country activities; the founding of an Environment Department at the World Bank also prompted change. Within a few years other international agencies and major NGOs established similar environmental facilities.

By the time the UN Conference on Environment and Development was held in 1992 there had been a 'sea change' in attitudes towards environment. While support at Rio was more verbal than financially committed there had been a large shift in attitudes since the 1972 Stockholm Conference, from suspicion about environmental care to recognition that environment and development interests were interrelated and vital (Colby, 1991). Unfortunately, the Rio 'Earth Summit' was less successful in generating workable conventions and agreements (Athanasiou, 1997: 8).

1992 to the present

There has been a trend toward addressing issues, although attention has perhaps been focused too much on the threat of global warming. Social studies, business, economics and law have focused far more on environmental management during this period and specialist support disciplines have diversified and expanded.

realistically, reckon on the basis of 'business-as-usual' scenarios and a failure to change enough – i.e. human habits will alter little. Environmental managers will have to prompt and nurture change in governance, ethics, economics, public attitudes and so forth.

An environmentally sensitive approach to development

Environmentally sensitive development could be defined as: use of what Nature provides to the optimum, and maintenance of that use indefinitely, avoiding ecological or social breakdown, in order to maximise human well-being, security and adaptability (see Figure 1.1). This demands high-quality management of the environment and human institutions, and the ability to recognise and avoid, mitigate or adapt to socio-economic and physical challenges and gradual unfavourable changes. Environmental management goes beyond seeking sustainable development; it also promotes improvement of human adaptability, the recognition and reduction of threats to people, biota and the physical environment, and rehabilitation of degraded ecosystems.

There is widespread complacency in rich and poor nations; many assume that current living standards, patterns of governance and technological progress will continue and hopefully improve under 'business-as-usual' conditions. That is understandable, but unwise, given that few nations have had more than 150 years without serious famine, and even fewer without large fatalities to epidemic diseases or warfare. Furthermore, the last 200 years has been one of the most climatically favourable periods of the last 10,000 years of marked changes and often inclement environment. There has been no major global catastrophe during recorded history, yet even over the last 500,000 years there have been mega-eruptions and other catastrophic events. Humans are more numerous than ever before and they are upsetting their environment, adding anthropogenic global changes to natural threats. Though it appears that there has been huge progress there is only a thin veneer of technology and governance protecting humans from disaster; people expect scientific and economic progress will continue without too much investment on their part and that there is no risk of breakdown. The crucial qualities, which in the past ensured human survival, are intelligence and adaptability, but many people

Figure 1.1 *Severely eroded valley in the High Atlas Mountains (Morocco). The area was forested as recently as 20 years ago; failure to manage land degradation has led to loss of forest and soil and a reduction of streamflow.*

Source: Author, 1998.

today have lost those skills. If technology and governance cannot compensate modern humankind will be far less able to withstand change than their predecessors.

The concept of environmental management

All individuals are to some extent involved in environmental management because their activities ultimately have an impact. Some individuals are more actively involved in resource use and interact more with Nature: fishermen, pastoralists, special interest groups, academics, applied researchers, administrators, government advisers and so on. So there are many different levels of environmental management. Environmental managers may deal with how to best tap resources; others are auditors, educators, impact assessors, or they may operate in a host of other ways. Often they are familiar with a specific sector, region or ecosystem – for example, dealing with tropical forests, coastal zones, mountain areas, petrochemical production and tourism. They may have sufficient powers to implement change, or have to consult others and lobby to try and achieve anything, or they may just be advisers. More and more those promoting or managing development are answerable to law and public scrutiny, and so consult specialists and encourage citizen participation. In countries with a tradition of open government and public access to information there has been more progress with this; elsewhere, the pace has been varied and the approach may be much less inclusive and transparent (Castro, 1993).

Environmental managers generally seek to understand the structure and function of the environment, the way humans relate to their physical surroundings and to one another. They then try to monitor change, predict future developments and try to ensure there is maximum human benefit and minimal environmental damage. Within larger businesses and government organisations, the environmental manager is generally a mid-level executive who oversees various specialists who collect data, monitor and research; he or she then takes the data to advise more senior management and develop strategic policy. An increasing number of companies or organisations maintain or hire in teams to establish and maintain an environmental management system (EMS), which helps develop, implement and review environmental policy. There are environmental managers who focus on understanding what is happening, monitoring for critical thresholds – points beyond which further change presents challenges – auditing performance and conditions. Others are more concerned with predicting the future, or policy making, or enforcement of environmental protection, or negotiating agreements, or establishing workable procedures.

There is a diversity of environmental managers, but most share some or all of the following characteristics. They make deliberate efforts to steer the development process to take advantage of opportunities; ensure no critical limits are exceeded; try to avoid threats; mitigate problems; and prepare people for unavoidable difficulties by improving adaptability and resilience (Erickson and King, 1999). Thompson (2002: 5) suggested environmental management was '... the system that anticipates and avoids or solves environmental and resource conservation problems ...'. Alternatively, it can be seen as a process concerned with human–environment intereactions, which seeks to identify what is environmentally desirable; what are the physical, economic, social and technological constraints to achieving it; and what are the most feasible options (El-Kholy, 2001: 15). Environmental management is concerned with meeting and improving provision for human needs and demands on a sustainable basis with minimal damage to Nature. Since the late 1980s a core concept and a goal of environmental management, especially in relation to development, has been sustainable development.

Environmental management is still relatively young and is rapidly evolving, so judging how successful it has been and in what ways it should be 'tuned' to better serve the quest for development is a challenge. As mentioned a paragraph or so ago, environmental management has to cope with natural threats and problems caused by human activity, it has to do this in a world where nature is being degraded, and it has to support livelihoods and steer these to ensure sustainable development. Environmental issues are so intertwined with socio-economic issues that it has to be sensitive to them, especially in poor developing countries – environmental management is '... of a single piece with survival and justice ...' (Athanasiou, 1997: 15). Environmental management has to do all this in a 'real world' where:

- Greed, corruption and foolishness conspire to hinder.
- Poverty and growing populations limit the options available.

- Knowledge and technical skills are still too limited.
- Increasing numbers of people demand more and more material benefits.
- The time available to make real progress is limited.
- Natural and human disasters may happen.

Environment and development issues are more and more transboundary and often have to be dealt with on a global scale. Law, governance, the sciences and management are still adapting to address this. In the past scientists have been able to thoroughly research problems and then suggest solutions, but increasingly advice has to be offered before adequate data and knowledge are available, because delay risks the challenge developing into a costly or uncontrollable problem. Environmental management faces unexpected and rapid changes and also insidious problems which develop so slowly and/or covertly that they get overlooked. For some problems novel intergenerational approaches may be required – something new to human development.

Environmental managers do not just cope with challenges, they have to model and monitor to gain sufficient knowledge and try to get early warnings, if they are to have adequate chance of success. Recognising a problem is one thing, reliably identifying causation is another. Problems may have complex indirect and cumulative causes – a number of unrelated factors suddenly conspire to make trouble, or a process develops a positive or negative feedback which (respectively) quickly accelerates or slows developments. Environmental management also has to cope with changes of fashion and economic conditions, new technological capabilities, alterations of attitudes, changed social capital, social values, skills, confidence and so on. Responses frequently have to be multi-disciplinary, and environmental managers must determine how people will be affected and react, and weigh what is the best way to cope.

Before the 1980s environmental management was mainly practised by those with a science background – ecologists, pollution specialists and so on – or concerned with monitoring and enforcement. There has been a very marked broadening out during the last decade or so, to the extent that environmental management is now more than half staffed by those with social studies skills. No single environmental manager can cope with all issues, and in academic institutions, funding agencies and large organisations, sociologists, anthropologists, economists, geographers, physical scientists, lawyers, planners and engineers increasingly come together in environmental management departments.

Very diverse stakeholders are involved in development and environmental management and often are little aware of each other. Urban citizenry are often ignorant of rural areas – the source of their water, food and energy; academics and scientists have to struggle to communicate effectively with administrators and the general public, special interest groups, NGOs, powerful multinational companies and politicians who are keen not to make a wrong move. Many problems must be solved in the face of apathy or forceful opposition. Politicians, some NGOs, lobby groups and individual 'gurus' may get away with advocacy, but environmental managers have to 'produce the goods' and perfect and carry through policies, programmes and projects which work.

Environmental management is unlikely to be wholly dispassionate – the practitioner's politics, ethics and outlook inevitably have an influence. Political and economic pressures often drive environmental management issues, and there are situations when it may be expeditious to work with, or even manipulate, ginger groups (groups which lobby for change), media and citizens. Environmental management can be highly political. Caution is also needed to prevent environmental managers becoming bureaucrats who are out of touch with best practice, popular feeling and global needs. It is valuable to heed the concerns of local people and to make good use of their expertise; but not to be parochial, keen only to meet local, regional and national wishes, and blinkered to wider needs.

Although developing countries vary a great deal in their type of government, historical fortunes, degree or poverty, natural resource endowment and so on, many have a tropical or subtropical location. In the past this led many environmental determinists to blame lack of development on topicality. While it is now accepted that environmental conditions do not *determine* development success, attempts to 'develop' in poor, tropical countries often face similar socio-economic and physical hindrances:

- High temperatures, erratic and sometimes extreme weather conditions and intense solar radiation frequently conspire to make water supply a challenge.
- Deeply weathered soils tend to rapidly lose organic carbon and are commonly poor in plant nutrients and rich in aluminium or iron compounds. These can easily suffer degradation.
- There are relatively complex biota, some of which are crop pests or disease vectors, with no marked cold season to control insects and weeds.
- Pest organisms develop rapidly thanks to warm conditions.
- Development efforts have often taken the form of large-scale, costly and inflexible projects, frequently implemented in a hurry by expatriate consultants on short-term assignments. Also, there may be no recurrent funding.
- The benchmarks used to judge the progress and success of development have commonly been inappropriate, paying attention to economic or engineering criteria, and giving too little attention to environmental and social issues.
- The developers may have inadequate local knowledge because they are frequently expatriates or overseas-trained city folk, and so tend to be insensitive to poverty, socio-cultural issues and environment.
- Development is conducted against the clock: in order to achieve goals before a government runs out of its term of office, or so that contractors can get early completion bonuses, or to cut costs, or because there is a genuine sense of haste to achieve development.
- Hindsight experience is not adequately shared because it is restricted to limited-circulation consultancy reports or academic journals which poor countries cannot afford to access; also, post-development appraisals are

seldom satisfactory because there is scarce funding, or those involved do not want to highlight 'shortcomings'.

- The art of precautionary planning has evolved quite recently and is still being adapted to real-world conditions; consequently, it is easily side-stepped, or is applied too late to select the best development option, or it is misused or neglected, or just lacks the power to identify impacts well enough.
- Funding may be allocated to developments which satisfy powerful special interest groups or bolster national prestige, rather than produce maximum utility with minimum environmental and social impacts.
- Expatriate and indigenous 'middle-class' experts and decision makers are reluctant to 'rough it' in the countryside, leading to slow responses to problems and patchy oversight.
- Civil unrest and lack of investment conspire to prevent collection and maintenance of adequate baseline data, or servicing of infrastructure and law and order.

Furthermore, the countries in which environmental managers operate frequently share similar challenges:

- Dependency – reliance on foreign inputs and commonly a need to adapt technology and techniques to developing country situations
- Globalisation – powerful pressures from overseas
- Debt burden and general poverty
- Falling national incomes because of declining commodity prices
- Breakdown of traditional resource management strategies and disempowerment of people (in particular loss of access to common resources)
- Inadequate governance, leading to corruption, unrest and law-enforcement problems, and powerful special interest groups
- Often high population density and rapid demographic growth
- Rapid social change
- Limited powers with which to negotiate with and influence developed countries and world bodies
- A legacy of history – inappropriate borders, unsuitable infrastructure, legislative 'hangovers' from colonial days, dependency, neo-colonial pressures and so on
- Centralised bureaucracy, which may have little clear oversight or much interest in more remote areas and poor people.

Environmentalism, environmental management, environmental ethics, environmental legislation and techniques for monitoring and forecasting have mainly originated in the Western 'liberal democracies', and have had to adapt, or are still being adapted, to fit developing countries (Lafferty and Meadowcroft, 1996; Gupta and Asher, 1998). Given that most of this has taken place in the last 30 years or so, there has been much progress. However, environmental management tools and methodology are still evolving, and the database of environmental and social knowledge for many developing countries is still woefully inadequate.

Concluding points

- Environmental management has been evolving and spreading for over 30 years. It has still to be adequately adapted to developing country conditions.
- Environmental management supports a proactive approach to development and better co-ordination of stakeholders' interests.
- Sustainable development is a key concept within environmental management.
- Science and social science are involved in environmental management and must work closely with development management.
- Without environmental management, development is unlikely to be sustainable and people will be vulnerable to disasters.

Further reading

Adams, W.M. (2001) *Green Development: environment and sustainability in the third world* (2nd edn.). London, Routledge (1st edn. 1991). [Excellent introduction to environment and development, sustainable development and the background of environmentalism.]

Allen, T. and Thomas, A.R. (eds) (2000) *Poverty and Development into the 21st Century*. Oxford University Press, Oxford. [Chapters introduce development and also environmental degradation issues.]

O'Riordan, T. (ed.) (1995) *Environmental Science for Environmental Management*. Longman, Harlow. [Good introduction to the physical aspects of environmental management, but with limited focus on developing countries.]

World Development Institute (2002) *World Development Indicators 2002*. Available on CD-ROM by faxing: 1-800-645-7247 or available online at http://www.worldbank.org/data/wdi2002/index.htm. [Indicators for measuring development and sustainable development.]

McNeill, J.R. (2000) *Something New Under the Sun: an environmental history of the twentieth century*. W.W. Norton & Co., New York (NY). Penguin edition available. [Readable, thought-provoking historical approach to environment and development.]

Power, M. (2003) *Rethinking Development Geographies*. Routledge, London. [Chapters 1 and 2 are a stimulating introduction to development.]

Pepper, D. (1984) *The Roots of Modern Environmentalism*. London, Croom Helm. [Introduction to the evolution of environmental thought and environmentalism.]

World Commission on Environment and Development (1987) *Our Common Future* (the Brundtland Report). Oxford University Press, Oxford. [This book played a key role in establishing sustainable development.]

Websites

Institute of Ecology and Environmental Management: http://www.ieem.org.uk

International Network for Environmental Management (world federation of national associations for environmental management): http://www.inem.org

National Association for Environmental Management: http://www.naem.org

Centre for Sustainable and Environmental Management: http://www.csem.org.uk/

Centre for Science and Environment (Delhi): http://www.cseindia.org

World Resources Institute: http://www.wri.org/

(All accessed January 2004)

2 Environmental management and developing countries

It cannot be in the interests of our society for debate about such a vital issue as the environment to be based more on myth than on truth.

(Lomborg, 2001: 31)

Key chapter points

- This chapter examines the evolution of environmental management.
- Key environmental management issues are introduced, especially: the polluter pays principle; the precautionary principle; and sustainable development.
- Environmental management is spreading to developing countries and is being developed by them, and is evolving rapidly. Not surprisingly, it is often not sufficiently adapted to developing country needs.
- Environmental management has to cope with inadequate data and knowledge gaps yet make prompt decisions that can be relied upon. It must also identify and deal with a wide range of stakeholders.

On average, modern humans have nine times more income per capita than their ancestors had in AD 1500, and four times as much as people obtained in 1900. Despite gross inequalities in the distribution of this income growth and considerable poverty in many countries economic development must count as an achievement. However, the achievement has come at a price. The social price, in the form of people enslaved, exploited or killed, has been enormous. So has been the environmental price (McNeill, 2000: 7).

The evolution of environmental management

Environmental management in the 1970s was largely technocratic and problem solving, the ethos being 'Trust me, I'm a professional', providing practical assistance to state officials in developed countries, and paying limited attention to social issues. Effectively it serviced state administration of the environment, undertaken on behalf of citizens (Bryant and Wilson, 1998: 321–2). It had by that point evolved discrete sectors: fisheries management, wildlife management, mining, agriculture, pollution control and so on. And it was largely top–down in approach, implementing and enforcing environmental policies in the main by coercion: laws, fines and closure for breaches of regulation. The last 20 years or so have seen environmental management shift to accept public accountability and consultation, and often abandonment of coercion in favour of reward and education (Martin, 2002). Once technical knowledge and skills were enough, now ethics, quality standards, codes of conduct and transparency are increasingly important.

The social sciences as well as the environmental sciences have a long tradition of exploring the human–environment relationship (Barry, 1999). Cultural ecology emerged in the 1930s as a school of ecological anthropology which explored the degree to which and manner in which culture and environment determine human fortunes. By the 1950s and 1960s it was seen by some as a better way than environmental determinism or environmental possibilism to study how human societies and cultures adapt through subsistence patterns to a given environment (Ellen, 1982: 53). Cultural ecologists argued that culture interacts with environment through processes of adaptation, which are affected by technological innovation, social change and so on. After the 1960s, interest in cultural ecology dwindled, and attention shifted to historical ecology and political ecology. Political ecology seeks to understand relationships between society and environment, or more specifically the power relationships – valuable for those wishing to understand and control environmental stakeholders (Bryant and Bailey, 1997; Keil et al., 1998:1; Stott and Sullivan, 2000; Zimmerer and Bassett, 2003). In the 1970s human ecology was seen by some to be the way to explore the human–environment relationship, but it has been less used in the last 20 years, although Glaeser (1995) argued it could help integrate environmental management and agricultural development.

Currently, the political ecology approach is quite often applied by environmental managers seeking to identify the politically located ideas which influence how people relate to their environment (Atkinson, 1999; see *Journal of Political Ecology*, available online at http://www.library.arizona.edu/ej/jpe/jpeweb.html). For example, Davis (2001) published a political ecology view of famines, arguing that various cultures had better withstood natural disasters before colonial intervention. Others have applied political ecology to assessing why people cause environmental damage. A recent development has been to research 'hidden voices' of people – a liberation ecology exploration of the human–environment relationship; there has also been interest in radical ecology and postmodern environmentalism (Merchant, 1992; Sarkar, 1999; Peet and Watts, 2004). Liberation ecology draws on social theory to offer political and economic explanations for many of the world's environmental problems, often presenting the environment as an area of struggle.

Environmental management is no longer just concerned with physical and largely quantitative data – it now deals with historical data, policy formulation, social capital and institutional issues, qualitative information from focus groups and questionnaires, advice from political ecologists, the findings of cultural ecology, anthropology, and much more. On the whole, it has become more co-ordinatorial and participatory and much more integrative. Modern environmental management also tends toward less compartmentalisation, more encouragement and support than enforcement, and often a 'bottom–up' approach (there is perhaps a fixation with participation and empowerment at the moment). There is also a shift toward continuous improvement and environmental stewardship, rather than command and control and solution of problems after they occur (Crognale, 1999). The dissemination from developed countries is still underway so environmental management is evolving to suit new situations (see Sawhney, 2004 for a review of how environmental management in India has changed over the last decade). Environmental concern and environmental management are influential: for example, they seem to have played key roles in *perestroika* (the liberalisation of the authoritarian centralised-state Soviet Union since the mid 1990s). Where once environmental management was authoritarian and centralised, the trend has been toward decentralised, often community-oriented approaches (Kapoor, 2001).

Colby (1991) presented five environment-development paradigms, which are not wholly discrete – there is much overlap. These have influenced the evolution of environmental management and each is characterised by different attitudes, capabilities and so forth. Colby provided a far more detailed discussion than the one that follows; however, this précis serves to outline the paradigms:

Frontier economics: an approach prevailing in most countries until the 1960s. Nature was treated as an infinite supply of resources to be used by humans, and a limitless sink for wastes. Managing the environment was more or less irrelevant because it was 'outside' economics. Technology was for improving human welfare and successfully stretching resources to improve crop yields, fish catches, energy supply and so on. The attitude to pollution was usually to clean up later, if forced to, or disperse it and forget it. Some developing countries still fall into this category.

Deep ecology: a worldview in many respects the opposite of frontier economics. It is a dark-green (deep-green) philosophy, with an ecocentric rather than anthropocentric outlook and a great diversity of supporters. It aims for harmony between humans and nature; it opposes the use of technology, and voices a wish to develop new ethics and development outlooks (Devall and Sessions, 1985).

Environmental protection: after the mid 1960s the frontier economics outlook weakened as pollution and biodiversity-loss problems became apparent, and as deep ecology supporters proselytised; it was seen to be necessary to make trade-offs between development and environmental protection. Tools like environmental impact assessment (EIA) were developed, and remedial measures were promoted to counter environmental damage. The norm was to seek 'end-of-pipe' pollution treatment or more-or-less managed dispersal. Various environmental agencies were created, often with poor co-ordination between them.

Resource management: this was supported by the publication of *The Limits to Growth* (Meadows et al., 1972). There were fears development would outstrip natural limits and cause disaster. Various strategies were published to counter the threat. One was to consider all resources and development plans, and draw up a system of national environmental accounts – essentially, accounting applied to natural resources and the environment – to ensure development lay within 'the limits'. The political economy and practical concerns of environmental management in developing countries are quite different from those in richer nations, and initially environmental protection was seen to delay and hinder development (Boyce, 2002). By the late 1980s concern for the environment was no longer greeted with open hostility.

Eco-development: appeared in the early 1980s (Riddell, 1981; Glaeser, 1984) and emphasised the need to restructure society and economics to ensure that development worked with, rather than against, Nature. The emphasis was on qualitative development rather than economic growth and on an awareness of the need for sustainability. Fields like industrial ecology, agro-ecology and ecological engineering appeared. Today environmental management is more forward-looking (proactive) and demands longer-term management of adaptability and resilience and efforts to reduce the risk of 'surprises' caused by inadvertently crossing a crucial environmental threshold or by natural disasters. One of the key roles of environmental management is to watch for and warn about critical environmental and socio-economic thresholds; consequently, predictive tools and indicators, which aid monitoring, are important (see Box 2.1). Monitoring and proactive impact assessment can be costly so environmental managers must select strategies that produce adequate information without exceeding their budget or taking too long (for coverage of the costs of environmental modelling see: Russel, 2003; environmental management tools are reviewed in Chapter 9).

Environmental management can be subdivided into several fields, including:

- proactive collection of information – environmental assessment; forecasting; 'hindcasting' (using history or palaeoecology); scenario prediction; risk assessment, etc. (Lave, 1987; Covello and Merkhofer, 1993; Norton et al., 1996; for further information see specialist journals such as *Human and Ecological Risk Assessment*);
- corporate environmental management activities: company environmental management, industrial ecology and related fields;
- environmental management systems and quality issues;
- environmental standards, enforcement and legislation;
- environment and development institutions and ethics;
- sustainable development issues;
- environmental economics: paying for environmental management, using economics to improve or enforce environmental management;
- environmental planning and management;
- stakeholders involved in environmental management;
- environmental perceptions and education.

Key issues and approaches in environmental management

Environmental management is developing so rapidly that selecting key issues is not easy. I have already mentioned that efforts are being made to improve

Box 2.1

Environmental management tools and roles (some have more than one role):

Future situation assessment

- Forecasting, scenario prediction, brainstorming, modelling
- Hindcasting – use of historical records and palaeoenvironmental data to identify likely future issues and to see how people have reacted in the past
- Backcasting – identifying future goals and then planning how to reach them
- Environmental impact assessment, social impact assessment, technology impact assessment, Delphi technique
- Risk assessment, hazard assessment
- Life cycle assessment
- Environmental modelling, socio-economic modelling.

Data gathering and early warning

- Researching structure and function of environment and organisms
- Identifying thresholds – points at which a problem develops that can be monitored for
- Geographical information systems (GIS) and remote sensing – storing, sorting, presenting and collecting data
- Social surveys (e.g. focus groups) to assess attitudes and collect social data
- Modelling – simplify situation to aid understanding or prediction
- Rapid rural appraisal (RRA) and participatory rural appraisal (PRA) – multidisciplinary and rapid data collection under developing country conditions
- Monitoring for environmental and socio-economic thresholds
- Being alert to warnings, lobbying on Internet from NGOs or concerned individuals.

Education and implementation

- Educating citizens, business and government officials about environmental management/ sustainable development; media – TV and information technology – are especially important.
- Public relations skills/tools
- Goal selection/implementation tools
- Strategic overview – strategic environmental assessment; strategic environmental management
- Policy- and decision-making tools
- Institution building to support environmental management and sustainable development – establishing bodies, social capital, etc.
- Fund raising/paying for environmental management and sustainable development
- Enforcement of or encouraging support for policies.

Stocktaking

- Site assessment
- Cost-benefit assessment
- Benchmarking – critical reference points against which judgements can be made

- Eco-auditing
- GIS
- Environmental accounting/valuation
- Environmental management systems
- Sustainable development auditing
- Project, programme, policy appraisal and evaluation – learning from hindsight.

Note: The term 'tool' is not precise; some tools can be used for a range of things: data collection, monitoring, appraisal. There is a possibility that expert systems could be developed for some of the above activities. These are computer programs prepared to guide relatively unskilled staff so that they can conduct quite complex tasks. Most of these programs improve with use – 'get smarter'.

understanding of human–environment relations; environmental economics is fast evolving; so is environmental justice and environmental ethics. The 'greening' of economics has coincided with quite broad acceptance by rich and poor countries that environmental issues are important, and not a hindrance or luxury (Pearce and Barbier, 2002). That sea-change took place somewhere between 1972 and 1992. Tools for effective environmental accounting are still far from perfected but there has been great progress. Environmental law has also been evolving fast, and also taxation, but these are also inadequate.

The argument can be made that what cannot be measured, cannot be managed. The recognition of useful indicators, and the development of effective monitoring, forecasting, and management decision-making techniques, are vital (Jeffrey and Madden, 1991) (see Box 2.2). Environmental management also demands skill in reading the public mood, so as to get support and establish what can be achieved. Partly related to the former point; discrete issues are more likely to attract interest than slow-onset and insidious developments, even if these are seriously threatening. It also helps if environmental management can point to clear benefits from its actions and not just flag threats (El-Kholy, 2000).

Box 2.2

A clarification of the roles of environmental management frameworks, EMSs and environmental management tools

- Think of a company or other body, or a sectoral activity (e.g. tourism) as an 'aeroplane'.
- An environmental management framework provides guidelines for planning the 'flightpath' to sustainability.
- An EMS is the 'flight manual' and 'checklists' needed to handle the 'aeroplane' (i.e. activity) in line with the framework.
- Various tools of environmental management (EIA, etc.) are the 'instruments' needed to monitor flight progress, and to steer it on track.

Source: Based partly on discussion in Robért (2000: 253)

The polluter pays principle

Over the last 30 years or so there has been a shift from 'develop now, clean up later' to 'avoid causing problems'. During that period there has also been a change from the burden of problems being borne by those affected, to it being shouldered by the public in general, but wherever possible making the 'polluter pay'. If forced to bear the costs for problems the potential polluter will be less likely to cause them; also, bystanders, consumers and workers do not pay for others' mistakes.

Penalties for pollution are still often hard to enforce and relatively light, especially in developing countries; consequently, organisations motivated by profit may be tempted to try to get away with sometimes getting caught and paying limited damages. Penalties for infringements thus have a weakness. In an ideal world environmental managers would educate and motivate people and organisations to genuinely seek to avoid polluting. In developed countries that has started to happen.

Sometimes environmental damage has become evident only years after it has been done; meantime, the body responsible has closed down or it is too late to use law to claim damages. The polluter pays approach seeks to make it difficult for responsible parties to escape damages and ensure that the penalty is enough to deter. If control is through licensing the applicant must convince the authorities that difficulties will not arise (with a safe margin of error); and for this risk assessment and impact assessment are useful.

The polluter pays principle should lead actors to internalise costs of their activities, which tends to increase business efficiency. It is widely seen as a 'twin' of the precautionary principle (see next section). Impetus to adopt the polluter pays principle has been given by disasters like that of Bhopal in 1984 (see Chapter 7). Also, the development of eco-efficiency has further prompted adoption of the polluter pays principle – because it can enable wastes to become useful by-products, improving profits.

The precautionary principle

The precautionary principle has no precise definition; it has been described as 'institutionalised caution', and is constructed around the goal of preventing, rather than reacting to, environmental harm (Applegate, 2000). The precautionary principle has (according to Kriebel et al., 2001) four central components:

- taking preventative action in the face of uncertainty;
- shifting the burden of proof to the proponents of a development;
- exploring a wide range of alternatives to try and avoid unwanted impacts;
- increasing public participation in decision making.

Acceptance of the precautionary principle in environment, health, economic and other policy implementation means that regulatory action is likely to precede full scientific certainty about an issue – lack of evidence is seen as no reason for inaction (Harremoës et al., 2002). Consequently it risks costs which may not be justified, and accusations of 'crying wolf' or delaying development. The precautionary principle is widely accepted in European and international law, but is less well established in the USA. Because there is no universal agreement on a definition its status is one of a broad approach, rather than a firm and precise principle of law. The 1992 Rio Declaration (Principle 15) of the United Nations (UN) Conference on Environment and Development (the Earth Summit) urged widespread use of the precautionary principle (O'Riordan and Cameron, 1994).

In the last few years there have been a lot of appraisals of the use of the precautionary principle, and some suggest it be discarded for all but general policies (Keeney and von Winterfeldt, 2001); although most accept it is valuable when serious and possibly irreversible, impacts are likely. A precautionary principle approach is useful for social development as well as environmental issues. In developing countries poor people have little in the way of security to fall back on if things like land reform or agricultural innovation fail – development efforts have to be right first time or some preparations have to be ready to give aid if there are problems. There is also the question of how much a society can afford to pay to support a precautionary principle approach. In practice – for example, in trade–environment disputes – its application is usually triggered by a risk assessment.

Although the USA lagged behind some other countries and international law in accepting the precautionary principle, it has nevertheless played a pivotal role in developing some key tools which support it, such as environmental impact assessment (EIA) and social impact assessment (SIA). Such impact assessment effectively forces developers to 'look before they leap', and if problems are anticipated, delay acting until there can be effective avoidance or mitigation. Applegate (2000: 421) described this as a 'speed bump' to warn of problems and if need be prompt slowing down to deal with them. The World Bank has started to shift from impact assessment of already identified developments, to using assessment (as it should be) to influence the identification and formulation of projects, programmes and policies in a much more strategic and proactive way (Goodland, 2000). The USA has also promoted legislation that seeks to ensure one of the things the precautionary principle seeks – to ensure margins of safety when designing technology, certifying drugs or pesticides, zoning areas at risk from avalanche or other disasters, setting standards for pollution measurement, and so on. Although a desirable goal, in many fields, margins of safety are far from reliably established, so enforcing them can be problematic.

The precautionary principle demands 'upstream thinking', looking for underlying causes of problems, rather than fixing on symptoms and *backcasting* – that is, visualising a desirable future scenario, identifying likely constraints, and then deciding what must be done now to move towards such a situation. Backcasting must not be confused with hindcasting, which is using historical and palaeoecological evidence to understand past events and trace human responses to help forecasting and assess probable reactions to future problems. Forecasting starts from today's situation and projects present solutions and responses into the future (Robért, 2000).

Lomborg (2001: 348) warned that the precautionary principle might be undesirable if it encouraged pessimism, which causes planners to abandon proposals, rather than going ahead and building with a margin of safety. He also warned that funds spent because of unsound use of the precautionary principle could mean less to spend on other things. There is also a possibility that environment and development assumptions are incorrect and could prompt inappropriate policies.

All of this presents environmental managers with a dilemma: there is often a need to act before knowledge and proof is available. That means risky decisions may have to be made. Politicians, financiers and most professionals are reluctant to take such risks, which could all too easily lead to loss of public trust. So, whenever possible, efforts are made to find 'win-win' paths; i.e., paths which pay off beneficially even if predictions prove to be wrong. Sometimes the payoff is direct and sometimes a useful opportunity is created. For example, a carbon-sink forest will have amenity, conservation and timber value even if global warning proves a false alarm (Karagozoglu and Lindell, 2000).

Phillimore and Davidson (2002) provided a fascinating case study of the application of the precautionary principle – the 'millennium bug' (Y2K) experience. They asked whether, given the minimal disruption that actually occurred, the precautionary expenditure of huge sums of money to address fears were worthwhile. Little research has been done on the misjudgement of the Y2K threat (Ravetz, 2000a). Another interesting aspect of Y2K was that bodies that tend to reject a precautionary principle approach spent huge sums using it. Global expenditure on Y2K probably cost over US$580 billion – it may well have been cheaper to do nothing and fix difficulties as they arose, and seems to have been the most expensive planning mistake since 1945. Organisations seem to have acted in a precautionary way because they believed that there were effective technical solutions; also, it was a discrete issue, they were clearly told what might happen and the computer industry is powerful and persuasive. Advocacy for many environmental management problems is likely to be less persuasive and an issue is unlikely to be so well defined. Another problem with the precautionary principle is that it can be anti-democratic, because it demands expenditure before a law or regulation has been broken or damage done, without the state or anyone else necessarily proving there is a problem.

Applegate (2000: 439) recognised the challenges the precautionary principle seeks to address and argued it had value for '... preventing injury under conditions of uncertainty ...'. The adoption of precautionary principle approaches is patchy; while governments may prepare against perceived threats posed by non-sudden events, studies suggest that, even if they are known to be likely, sudden events – floods, earthquakes and suchlike – tend to be dealt with after they have happened (Dery, 1997). Reviews of precautionary principle applications can be found in the *Journal of Risk Research* or *Journal of Environmental Law*.

Sustainable development

While the concept of sustainable development dates back to around 1970, wider interest in sustainable development was prompted by the limits-to-growth writings of the Club of Rome in 1972. Sustainable development offered a way to heed limits *and* develop, which seemed preferable to 'zero growth' (Meadows et al., 1972; Colombo, 2001). The Brundtland Report (World Commission on Environment and

Development, 1987: 40) did much to establish the concept and offered a useful definition: 'to meet the needs and aspirations of the present without compromising the ability to meet those of the future'. However, that definition is not precise and is but one of a huge number.

Some see sustainable development as a goal, others as a paradigm shift, or as a guide-rail for development. Clarke (2002) noted that the premise of sustainable development was that conservation and development are compatible because healthy communities depend on healthy environments, and that it had three goals: economic growth, environmental protection, and the health and happiness of people. Supporters of sustainable development do not seek environmental quality in isolation from addressing social disintegration and poverty (Downs, 2000); the latter cause environmental problems so must be addressed. Tough environmental standards are not acceptable if they cause poverty. The question is, can these ambitious goals be achieved on a wide enough scale, in real world situations, and within environmental limits?

Many see sustainable development as a valuable theory or paradigm – like justice or liberty – but fear there are too many problems for it to be widely put into practice. The majority, however, are seeking to put sustainable development into practice. Pursuit of sustainable development is helping to integrate socio-economic and environmental management (Kirkpatrick and Lee, 1993). Although now a prime objective of environmental management, it is a challenge to find effective and workable sustainable development strategies and governance (see Figure 2.1). Such strategies will frequently overlap and interact, so it is vital to ensure that they do not interfere with each other, which requires both a local knowledge and strategic co-ordination, ultimately at global scale. There have to be supportive human institutions, which must be resilient and adaptable to meet unforeseen challenges; there must be adequate information about the past, the present and the future (environmental, social, technical and cultural); and willingness to make sustainable development work. Part of the struggle will be to spread ethics which value sustainable development, foster productive social interaction and make better use of knowledge. It is highly unlikely that all constraints and challenges will ever be fully assessed in advance, so resilience and adaptability are crucial to any strategy (Turner, 1988).

International agencies and business are promoting sustainable development. One of the key bodies, the UN Commission on Sustainable Development, was established in 1993 by the UN Economic and Social Council to follow up proposals made at the 1992 UN Conference on Environment and Development and promote sustainable development, including supporting the implementation of *Agenda 21*. However, the Commission lacks 'teeth' (Bigg, 1995). Other bodies promoting sustainable development include: the World Business Council for Sustainable Development (http://www.wbcsd.org); the Division for Sustainable Development, a secretariat serving the Commission for Sustainable Development of the UN Department of Economic and Social Affairs (http://www.un.org/esa/susdev/about); and the International Institute for Sustainable Development (http://www.iisd.ca).

The range of tools and approaches for the management and promotion of sustainable development are growing. Approaches which are already widely used

Figure 2.1 *Tea estate, Sri Lanka – sustainable development? This estate has been in production since the early 1920s. It yields an export crop with minimal inputs, especially pesticides; the tea bushes provide some slope protection cover to reduce soil erosion; at this altitude and on poor and sloping soil few other tropical crops will flourish. Green tourism has been added on, making use of existing access roads, to help ensure the profitability of tea production. How sustainable the estate is in the very long term is difficult to establish but it is better than most available alternatives.*

Source: Author, 2003.

include: strategic sustainable development (Nijkamp and Soeteman, 1988; Robért et al., 2002); integrated appraisal (Lee and Kirkpatrick, 2000); environmental management systems (see Chapter 9); a range of indicators for measuring progress toward sustainable development; and economically sustainable economic development (Desta, 1999). Women in many societies are active in agriculture, fuel-wood collection and gathering forest products, and in some societies women control family finances; those pursuing sustainable development must be aware of gender issues (Jacobsen, 1992).

Environmental management can support sustainable development by:

- identifying key issues
- identifying threats, opportunities and limits
- establishing feasible boundaries and strategies

- co-ordinating the diverse physical, biological and socio-economic issues and overseeing the activities of various stakeholders.

Command-and-control or voluntary approaches to environmental management?

In the past it was usual to adopt a command-and-control approach to environmental management, relying on fines, regulations, licensing, inspections, bans, etc. Increasingly, such enforcement is giving way to a more 'hands-off' voluntary and reward-based approach, although clearly for some dangerous activities there must be rigid rules and strict controls. The voluntary approach includes the adoption of environmental management systems (EMSs – see Chapter 9), eco-auditing, eco-labelling, life-cycle assessments (see later this chapter), and negotiated agreements (see Figure 2.2). In Europe a widely used EMS is the voluntary Eco-Management and Audit Scheme (EMAS), which seeks to encourage companies to take responsibility for environmental management. Other EMSs are certified by international bodies, and unsatisfactory conduct results in an organisation being de-certified – which is likely to raise their insurance premiums, and possibly tax, and is bad for public relations (Honkasalo, 1998). EMSs certified under the ISO 14000 series are often selected because companies and institutions see them as more 'internationally acceptable'.

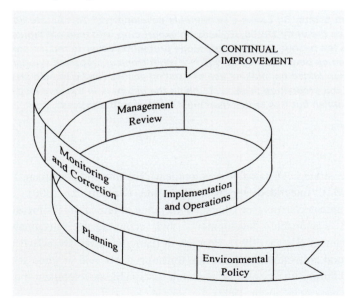

Figure 2.2 *The Environmental Management System (EMS) approach. The above cycle should ideally become more cone-shaped with time; successive cycles of EMS generally become easier because they are building on an existing base and have learnt from hindsight.*

Source: Redrawn and modified from Hunt and Johnson (1995) fig. 1.2, p. 6.

Honkasalo (1998) felt it unlikely that EMSs would eliminate the need for permits and other non-voluntary enforcement measures. Those will be needed for liability reasons and to ensure minimum standards – and are especially important for dangerous or nuisance activities. The world now uses a complex mix of voluntary self-regulation and enforcement, which may give adequate control over a compliant body's activities, but a comprehensive and strategic overview and other controls are also needed. EMSs are useful organisational or even regional administrative tools and help establish common benchmarks, but they are not a complete solution.

How environmental management is being pursued in developing countries

Developing countries have experienced a diversity of non-colonial, colonial and post-colonial regimes; these have included 'liberal democracies', 'free enterprise' dictatorships, socialist states, more-or-less constitutional monarchies, and many other governments. Often there has been a fair degree of what in Western democracies would be seen to be corruption and nepotism – which the North is by no means free from – frequent regime change, limited resources, vulnerability to outside pressure and an urge to 'develop'. These shape environmental management in the South. Empowering a country or local group to manage its environment is no guarantee it will seek to do so wisely or be allowed to by outsiders – romantics assume it will.

Colonial environmental management generally relied on expatriate expertise, seriously undervalued and ignored local knowledge and adopted a top–down approach. It placed concern for nature after revenue generation through export crops or mineral resources. Resource exploitation and resource accounting have thus often dominated environmental management in developing countries (Schramm and Warford, 1989). Some colonial environmental concern was driven by individuals who saw the need for stewardship, and to a lesser extent by 'romantic environmentalism'; but in those days there was very little pressure from NGOs, no politicisation of environmentalism and little public concern or media pressure. Indigenous peoples in particular were frequently cheated of their territory in one way or another and were seen as a nuisance, rather than source of useful wisdom (Crosby, 1987; Gadgil and Guha, 1992, 1995; Wall, 1994).

Some colonial governments established conservation areas, slope-protection forests, game laws and so forth. From the 1950s to 1980s anti-land degradation measures were often top–down and failed to win local people's acceptance. At independence a number of developing countries abandoned or neglected to enforce environmental management measures. The reasons range from resentment at what were seen as 'colonial restrictions', attention focused on social reform and economic development issues, and a view that environmental concern was a luxury that had to wait until enough economic progress had been made. The latter attitude was widely held until somewhere between 1972 and 1992. The cultures of developing countries that escaped colonialism may have had more freedom to shape their own environmental policies, and there are countries where non-Western attitudes have had more influence. For example, in some states Buddhism places value on preserving nature (Cooper and Palmer, 1998). There are plenty of cases where insensitive environmental management has been ignored or opposed. Numerous countries have countered soil erosion by tractor-ploughing terraces and planting trees. The costly measures have frequently failed – often the locals simply let their livestock graze and kill the trees, or even dig up the terraces. This is because they have no other land, see no need for the earthworks, or have no access to tractors or labour to maintain the measures. Environmental managers need to ensure development is 'in tune' with local needs, environment and social situation, and is understood by local people.

By 1992, environmental management was widely supported, at least in theory. However, poorer nations have limited funds and expertise, and are dependent upon developed countries and multinational companies. Pursuing optimum environmental management solutions may mean opposing outside influences, and that is seldom easy. Developing countries have become more aware of their environmental interests, obligations and aspirations and nowadays, thanks to NGOs and the media, have more power to bargain with and lobby richer countries and multinational companies. Sometimes the South can collectively bargain from a reasonably strong position because they hold important natural resources, can form cartels, or cast crucial votes in international meetings like the UN General Assembly. Global environmental problems and some opportunities involve both developing and developed countries, so the North has to liaise with and agree issues with poor countries (Miller, 1995).

Environmental management in developing countries faces the challenges encountered in richer nations, plus more limited funds, the need to combat poverty, hindrances like poor infrastructure; and risks that multinational companies and corrupt government will undermine efforts. Some developing countries have harsh environments which can pose particular environmental management and development problems, and often there is a diversity of ethnic groups with varying needs and aspirations (see Box 2.3) (Kasperson et al., 2003). Environmental management has to judge what is appropriate, and co-ordinate stakeholders. This may be done at a local, regional or global scale; the following list illustrates the diversity of levels at which environmental management can operate, and is partly based on Suriyakumaran (1979), and Bryant and Wilson (1998:325-326):

- *Grassroots (local or micro) level*: individual farmers, fisherfolk, herdsmen, slum-dwellers, and many others; family groups; single businesses. Individuals with a diversity of interests and little power are frequently seen as 'those to be managed' along with the environment. Only in the last decade or so has this group really been heeded and allowed to participate.
- *Sector level*: groups of companies or institutions with similar interests.
- *Regional level*: an environmental manager may use ecosystems, natural landscape units, or economic units – like watersheds, riverbasins, coastal zones, or other ecosystems, as 'regions'. A community-based watershed or comprehensive river basin approach has been adopted in many developing countries.

Box 2.3

Sensitive environments

Dry lands

There is little grazing or other agricultural activity in very arid areas, other than close to reliable streams or groundwater supplies – they are too dry. But, there is a long history of occupancy of areas which are semi-arid or seasonally dry. These have often experienced considerable increase in human and livestock numbers since the 1950s. There has also been widespread breakdown of traditional livelihoods and of coping strategies used in times of drought. Vegetation and soil damage is easy to initiate and difficult to remedy. Communications are often poor and people scattered, making provision of services difficult.

Highland areas

High altitude environments are harsh by virtue of their extreme and variable climate, often with marked daily temperature fluctuations. There are also high levels of ultraviolet radiation, thin soils, steep slopes prone to mass movements, thin atmosphere, frosts, exposure to winds and often exposure to pollution-bearing clouds and mist. Communications are generally poor, due to rugged terrain and bad weather. Once disturbed the slow-growing vegetation is difficult to re-establish and soil is easily lost. Slopes are then prone to mass movement – landslides and mudslides. Some highland soils are frozen for part of the year, most are thin and infertile. Where land is more level soils tend to be waterlogged. Where there is snowfall road communications can be difficult for much of the year and frost heave plays havoc with road surfaces. Snowmelt can result in a sudden spring release of several months' accumulation of deposited pollutants. Forage plants and soils often tend to retain radioactive pollution and toxic compounds. High-latitude regions lie mainly in more affluent countries, and are everywhere fairly sparsely populated; however, in Tierra del Fuego and Arctic Asia development efforts have to contend with intense cold and short daylength for more than half the year.

Thin and infertile soil areas

Large areas of the world have little topsoil, or crusts and subsoil layers that impede drainage and plant roots and can channel water to form gullies and hollows. Sandy soils drain rapidly and lose nutrients and organic matter fast. Some soils have high levels of toxic compounds. Many tropical soils are rich in aluminium or iron and, being already acidic, are vulnerable to acidification. Fine-grained soils, such as dense clays, may drain poorly, cling to implements, become hard and very difficult to till, during dry weather and can hinder communications in wet weather if tracks are not surfaced. Fine-textured soils are vulnerable to salinisation if groundwater rises anywhere near the surface. There are loess soils, which easily suffer wind and water erosion, and acid-sulphate soils, which become toxic to vegetation if drained, and peats which shrink or even ignite and burn below ground.

Wetlands and floodlands

Areas which are permanently wet, such as marshes, swamps and seagrass meadows, and floodlands inundated for part of the year often support considerable human populations and rich biodiversity. These people have usually developed strategies for coping with shifting water levels and for winning a living; often the adaptation is a form of transhumance – seasonal migration to higher ground. These areas get degraded by polluted rivers and dried up by diversions of flows and large dams. A few wetlands have suffered as a consequence of government hostility – e.g. rivers supporting swamps in southern Iraq were diverted to dislocate the Marsh Arab population. Wetlands and floodlands, so often neglected and degraded, have the potential to support sustainable agriculture and aquaculture.

Coastal areas

Coastlands often attract human activity and some are very sensitive environments. Mangrove swamps are easily degraded and are often developed for aquaculture, housing, industrial and dock-facilities; they are also difficult to rehabilitate if polluted with oil spills. Many of the world's coral reefs are being degraded by various activities. Reef and mangrove swamp damage both result in loss of biodiversity and expose coastlands to greater risk of storm damage. Tourism development can cause considerable damage to coastal areas, and there is a likelihood that global warming will have particular impact on these areas in coming decades.

- *State/national level*: some states are liberal, others more repressive, some are poor and isolated, and others can wield power through cartels or for strategic reasons.
- *Global level*: global environmental change and hazards, globalisation pressures and trade-related issues have huge impact. If it suits big business and powerful governmental or international agencies they can focus money, skilled manpower and lobbying to influence any country. Not all business is grasping and wholly profit-driven; there are companies that operate with strong corporate social responsibility and keen environmental concern.
- *International bodies*: like UN agencies or the World Bank. Most are rooted in developed countries and may not be wholly neutral toward developing countries' interests but have resources and high standards. A few are underfunded or are bureaucratic, inefficient and even corrupt.
- *NGOs*: many international NGOs work in developing countries or have established subsidiary groups, and growing numbers are wholly indigenous. Thanks to the Internet, NGOs often liaise closely and can network worldwide. Such NGOs have considerable influence through 'whistle-blowing' to world media or relevant authorities, and may hire top environmental and legal experts (Princen and Finger, 1995).
- *Special interest groups*: tend to be less open than NGOs: indeed, they may be extremely secretive. These are individuals or groups able to exert pressures to get their own way. They may be a significant section of a state's voters who clamour to be appeased, an international group or a very powerful minority.
- *Public opinion*: many environmental management issues are the sum of mass individual choices. Public opinion is frequently a crucial factor in controlling an environmental problem. Public opinion is fickle, often irrational, sometimes altruistic, commonly selfish, but might be moulded by media and education.

No single environmental manager or unit can cover all levels and sectors; however, a well-known environmentalist adage offers sound advice: 'think globally, act locally'. A hierarchy of environmental management bodies can be effective. Even in developed countries environmental management must cope with uncertainty, and in developing countries there tend to be even more surprises. One solution which has been advocated is an adaptive environmental management approach – however, it can be slow and costly and yield variable results (Holling, 1978; Walters, 1996). Strategic environmental management could be another way to deal with the challenge. There are rather 'fuzzy' definitions of strategic environmental management; basically it is approaches that seek a longer-term and larger-scale overview. So far, the application of strategic environmental management in developing countries has been limited.

There are many possible approaches to environmental management in the following listing of characteristics, where the left ('insensitive') is broadly what has been dominant, and the right ('sensitive') is what appears to be evolving.

Insensitive	*Sensitive*
top–down	bottom–up
command/coercion	hands-off/voluntary/encouragement
technocratic	appropriate (low or intermediate technology)
compartmentalised	integrative (even holistic)
authoritarian	participatory
non-sustainable	sustainable
anthropocentric	more ecocentric
centralised	decentralised

Because Western academics value participatory approaches or empowerment, does not necessarily mean these are welcome or appropriate elsewhere. Most developing countries sincerely back environmental management and have made significant contributions to its development; but some do so mainly to satisfy aid donors, and a few are active only when forced by treaties, the lobbying of NGOs, media or citizen action groups. In some cases poorer countries can make a relatively fresh start because there is limited encumbrance of existing, possibly outdated environmental bodies or legislation and they are free to acquire the newest technology.

There are countries where colonial legislation was either always inadequate or has been neglected since independence because it is seen to be anachronistic. Special interest groups are frequently powerful in developing countries, and it is quite common for expatriate interests to exploit resources. Multinational companies are already very powerful and are likely to become more so; with budgets which dwarf many a poor country's economy they can easily get their way. In some cases it looks likely that business interests and aid agencies will be a vehicle for environmental management. On the global level developing countries may be rewarded with trade and aid deals in return for voting on various treaties in ways rich countries and large companies want. While citizenry may spontaneously lobby for environmental reforms, the reforms are often orchestrated by international NGOs or other bodies.

Civil servants and professionals are widely trained in developed nations, and these staffs are helping to spread environmental management. Funding bodies often insist on environmental 'soundness' before they will support developments. However, it is not good for environmental management to come only from the North to the South. Expansion and evolution of environmental management must also take place in the South. That has started to happen, and new approaches have appeared in developing countries and spread worldwide – in particular, appropriate and intermediate technology solutions. Another promising area is the discovery by anthropologists and archaeologists of traditional farming strategies in the South that could be spread more widely there, and possibly in richer countries. In some cases the ideas may come from the South and then be upgraded or adapted for use elsewhere in the North. There is also a rich store of biodiversity in developing countries, the raw material for biotechnology, which the North also needs and should be made to adequately pay for (see Chapter 5).

Rapid urban growth, accelerating industrialisation, widespread natural resources exploitation, and in many regions a breakdown of traditional livelihoods means stresses which have caused some to talk of a Southern environmental crisis. Over the last 50 years a good deal of development effort has proved inappropriate and ineffectual: some schemes have been large and inflexible; sometimes the management has been weak, or the planning and engineering has been faulty. At first this may have been partly the result of the temperate environment outlook of many experts, which was unsuitable for subtropical or tropical conditions. While many standards and techniques still come from the North, developers should by now be aware of environmental and social conditions in non-temperate areas. The spread of the Internet and modern telecommunications has helped overcome some of the problems – expertise can now be tapped remotely, communications are easier so information can be shared, and problems cannot be easily hidden.

There are widespread fears that global environmental change will cause problems, which may differentially affect rich and poor countries. There is a need for all nations to act together to face global environmental challenges. Large populations with wide eco-footprints are vulnerable to global changes and will be difficult to cater for if there are serious environmental or socio-economic upsets. In the past a major disaster, like an Indian famine, affected only a few million people – now even single urban conglomerations like Mexico City contain over 20 million. Before the present day most countries were largely self-sufficient in food and energy – today most depend on long supply chains. Developing countries import grain and manufactured goods from developed countries, and developed countries buy food and fuel from developing countries – the world's people are interdependent.

Concluding points

- Environmental management has been evolving for more than 30 years and is still undergoing rapid change. Not surprisingly it is applied in many different ways.

- A precautionary and proactive approach is vital if sustainable development is a serious goal.
- Environmental managers need to know how humans interrelate with their environment, and often draw on fields like political ecology to assist.

Further reading

Chiras, D., Reganold, J. and Owen, O. (2002) *Natural Resource Conservation: management for a sustainable future* (8th edn.). Prentice-Hall, New York (NY). [Good introduction to environmental studies and sustainable development.]

Porter, P. and Shepard, E. (1998) *A World of Difference: society, nature, development.* Guilford Publications, New York (NY). [Focuses on development and environment.]

Brenton, T. (1999) *The Greening of Machiavelli: the evolution of international environmental politics.* The Royal Institute of International Affairs, London. [History of international environmental negotiations relating to environmental co-operation.]

Bryant, R.L. and Bailey, S. (1997) *Third World Political Ecology.* Routledge, London. [Good coverage of political ecology and development.]

Stott, P. and Sullivan, S. (eds.) (2000) *Political Ecology: science, myth and power.* Arnold, London. [A political ecology focus on developing countries and environmental policies affecting them.]

Keil, R., Bell, D.V.J., Penz, P. and Fawcett, L. (eds.) (1998) *Political Ecology: global and local.* Routledge, London. [Case studies providing insight into contemporary environment and development issues from a political ecology standpoint.]

Websites

Precautionary principle

Outline and bibliography: http://www.biotech-info.net/precau

Critique of USA and EU usage (Heritage Foundation): http://www.heritage.org/research/

Critique (Social Issues Research Centre Oxford): http://www.sirc.org/articles/beware

Commercial viewpoint: http://edie.net/news/archive

EU Commission: http://european.eu.int/comm/dgs/heal

Practical example: http://www.jncc.gov.uk/marine/fish

Sustainable development

International Institute for Sustainable Development: http://www.iisd.org

Foundation for Sustainable Development (grassroots focus): http://www.fsdinternational.org

UN Department of Economic and Social Affairs – Division for Sustainable Development: http://www.un.org/esa/sisdev/csd

World Business Council for Sustainable Development: http://www.wbcsd.ch/

Centre for International Sustainable Development: http://www.cisddl.org/

WorldWideWeb Virtual Library: http://www.ulb.ac.be/ceese/meta/s

Sustainable Development Communications Network: http://www.sdnetwork.net/

Sustainable Development (journal – Wiley): http://www3,interscience.wiley.com.gi-bin/home/s346

The International Journal of Sustainable Development (journal - Parthenon Publishing): http://www.parthpub.com/sustdev/h

Sustainable Development International [online journal]: http://www.susdev.org/

Food and Agriculture Organisation (FAO) Sustainable Development Site: http://www.fao.org/sd/index-en.htm

Earth Summits

1972 Stockholm Earth Summit (UN Conference on the Human Environment): http://www.mylinkspage.com/earthsummit.html

1992 Rio Earth Summit (UN Conference of Environment and Development): http://www.mylinkspage.com/earthsummit.html; .un.org/geninfo/bp/enviro.html

2002 Johannesburg Earth Summit (UN Conference on Sustainable Development): http://www.mylinkspage.com/earthsummit.html; .earthsummit2002.org/; http://www.johannesburgsummit.org/

Environmental NGO

Consumers Association of Penang – developing-country-based environmental NGO: http://www.jeef.or.jp/EAST_ASIA/malaysia/CAP.htm

(All accessed February 2004)

Part II
Resource management issues by sector

3 Water, coastal and island resources

Key chapter points

- This chapter explores the problems caused by competition for water. Demand is increasing so supplies must be better managed and valued more highly. In the past water was sufficiently abundant to be relatively easily managed and 'engineered'. Generally the best supplies have been developed and in more and more cases what is available is less than optimal, having often been damaged.
- So far, water resources have mainly been developed by water managers with a relatively narrow outlook – usually concerned with economic and engineering goals. Environmental management applied to water resources promises to better integrate the needs of all stakeholders and give a more comprehensive overview.
- Water supply, especially in developing countries, has focused on quantity. There are calls for marked improvements in quality – better management of chemical and micro-organism content – in the early twenty-first century. That demands more investment and a more comprehensive approach to development. Solutions to water supply problems have so far relied on project engineering, especially large dams and large-scale transfers. These have been problematic; however, there are still such developments in progress.
- Coastal and island ecosystems, like river basins, come under intense human pressures and therefore stress. In many parts of the world coastal and island environments are being degraded badly. Without careful environmental management their use will not be sustainable.

Water

Fresh water is vital for the survival of humans and most other organisms, and little development can take place without adequate supplies. It is frequently a sustainable resource if managed well, although some water supplies are finite and no matter how carefully used have a finite life. To sustain water resources demands adequate environmental management; this is often lacking, resulting in degradation, and even temporary or permanent loss.

Excessive exploitation and pollutants can damage surface water and groundwater supplies. Pollution may be from industry, sewage or livestock waste, agrochemicals or warfare and may follow bushfires. Another cause is poorly managed soil and altered vegetation cover; these may lead to diminished and contaminated water supplies. The damage may be from poor land use leading to nutrient overenrichment and cloudy, sediment-contaminated water, which damages plankton, fish, shellfish and aquatic plants, and silts up channels, lakes and reservoirs. Thin and compacted soil can cause precipitation to run off without infiltrating – the result can be severe erosion, flash flooding and reduced groundwater recharge. Reduced groundwater recharge can cause springs and streams to dry up, wells to run dry and river flows to become more erratic. Floods are also caused by mismanagement of soils and vegetation and ill-judged drainage, channel improvement and construction, and can harm huge numbers of people. There are widespread signs of pollution and siltation of marine ecosystems, which can be traced to mismanagement of land and rivers.

Problems are often caused by water-borne sediment. Streams, rivers, channels and lakes choke with sediment, causing floodwaters to overflow banks and destroy bridges and other infrastructures. Reservoirs and tanks lose their storage capacity – even very large hydroelectric schemes are often ruined by silting, sometimes before they can pay back construction loans. Irrigation schemes, water filtration plants and sewerage systems fail, and navigation channels and harbours require expensive dredging. Many organisms in rivers, lakes and the sea are intolerant of increased sediment. Sediment pollution is hugely costly, often dangerous and usually unsightly, which can damage tourism.

Sediment is widely generated by land development with poor soil and water conservation, but is also generated by mining waste, overgrazing and inadequate range conservation, and often follows forest fires. Sediment is a lost resource for the areas eroded, but where deposition occurs there may sometimes be benefits from the fertile silt; however, some deposition damages farmland and infrastructure. Sediment transport is often episodic and can be very difficult to monitor: much is generated during storms, especially in the period after tillage before crops provide ground cover. More funds need to be found for monitoring sediment loads in streams and rivers and for improving land management to reduce sediment discharge.

Adequate soil and water conservation approaches, which reduce runoff and thus retain soil and moisture to sustain agriculture and recharge groundwater, should help reduce flooding and prevent silting. So, groundwater, surface water and soil management should not be conducted in isolation; the approach should be comprehensive and integrated (Thana and Biswas, 1990). 'Adequate water supply' means enough of a suitable quality, when it is wanted and where it is needed, and satisfactory disposal of any surplus. However, the world's precipitation and distribution of rivers and groundwater is far from spatially or seasonally well distributed. Precipitation may fluctuate as a consequence of natural climatic change, recurrent events like El Niño, and possibly anthropogenic climate changes. Large parts of the world have seasonal shortages; for example, most of India has about 80 per cent of its rainfall in the months of June–September (Swain, 2001). Human development must either adapt to uneven water availability or manipulate supplies through strategies like reservoir storage. Where temperatures are high and soils drain freely, which is often the case in developing countries, even a high annual rainfall may be inadequate, especially as populations increase and demand grows, so management is crucial.

Much water management has been dominated by engineering approaches, often on a large scale, and judged mainly by economic and technical criteria. Water resources must often support domestic supply, tourism, conservation, power generation, flood control, sanitation and health, navigation, agricultural development, pollution control and other demands. These demands often conflict unless carefully managed. Water management is often complex and politicised and depends on sound institutional developments – a comprehensive and integrated approach is therefore widely advocated. Adaptability is also important when developing resources in areas where inadequate data make it difficult to safely forecast, and everywhere unforeseen changes like global climate or sea-level rise could spell disaster for inflexible water development. Unfortunately, much water resources development in the South is not flexible. There is often a need to repay large loans, and recurrent funding for maintenance. Modifications and management are frequently inadequate. Engineering can rule out modification and not allow enough leeway to deal with the unforeseen. Inadequate expertise hinders adjustments, and the management may be weak.

Many environmentalists identify water supply as *the* environmental and social challenge of the twenty-first century. Yet it is fair to say most people have not realised this, and water is too often taken for granted: even after the First UN Conference on Water at Mar del Plata in 1977 there was surprisingly little interest. Chapter 18 of *Agenda 21* published at the Earth Summit in 1992 did pay some attention to water, but proposals were poorly formulated, and those hoping for a World Water Convention to focus attention and resources have so far been disappointed.

Postel (1992: 24) noted that the world was entering a new era, one that contrasts with earlier decades when damming rivers and drilling wells was relatively straightforward, and supplies were fairly easily 'engineered'. The coming generation will face more constraints: political, environmental and economic. Water will be in shorter supply and more likely to be competed for (see Figure 3.1). Already in many countries the best supplies are in use, so future demand will have to be met by less wasteful usage and from less than optimal sources. Water crosses borders above and below ground; rivers and lakes often form territorial boundaries, and the shared waters are often crucial for the nations concerned. Where rivers and lakes are shared by more than one country it often looks likely that demands will exceed what can be met by simple abstraction. Water management therefore has to effectively address transboundary issues and sharing, and explore recycling, reduction of wastage and alternative sources.

Since the late 1990s lobby groups have been trying to stimulate concern, notably the World Water Council established in 1996 as an international non-governmental organisation (NGO), not a United Nations (UN) body, to raise awareness and act as an international think-tank. Other groups include the Global Water Partnership

Figure 3.1 *Water-stressed and water-scarce countries (note the islands circled) – projections for AD 2025.*

(established 1996), and the World Commission for Water in the Twenty-First Century (Abu-Zeid, 1998). The latter body has called for:

- Holistic and systematic approaches and integrated management of water resources
- Full-cost pricing for water, and targeted subsidies for the poor
- Institutional, technical and financial innovations in water management.

The UN declared an International Drinking Water Supply and Sanitation Decade between 1981 and 1991 to try and improve the health situation in developing countries, with limited success, partly because of population growth raising demand. International Conferences on Water and the Environment were held in Dublin in 1992 and 1999, and others followed in Paris in 1998, and more recently – these have helped focus attention on water resource issues. In developing countries the International Water Management Institute (IWMI) works to improve management of freshwater resources, supporting studies and development of irrigation, river management and so on. In 1997 the UN published the *Comprehensive Assessment of Freshwater Resources of the World*, and the World Water Council has helped organise a number of World Water Forum meetings, the first in Marrakech in 1997, bringing together interested NGOs, UN agencies, aid donors and private bodies. The second Forum, held at The Hague in 2000, was one of the main water-focused meetings since Mar del Plata in 1977. However, authorities like Biswas (2001) warned there was a vital need to increase concern for water, especially to break away from just focusing on the present to look more to the future. An International Year of Fresh Water was declared in 2003 and emphasised supply and access, during which UNESCO published *The UN World Water Development Report 2003* (available online at http://www.unesco.org/water/wwap or http://www.berghornbook.com; accessed October 2003) which was the result of combined efforts by 23 UN agencies and some other bodies. This Report presents key issues and provides a global overview of the state of the world's freshwater resources.

Water sources

Freshwater supplies are mainly obtained from:

Surface runoff: either short-lived flows following precipitation, or more stable streamflow. The latter may still fluctuate a great deal and prompt developers to construct dams and storage reservoirs.

Groundwater: underground supplies held in water-storing rocks (aquifers). These may be renewed by precipitation and surface flows seeping into the ground, or by artificial recharge – and with good management the supplies may be sustained indefinitely. However, in some cases they have accumulated in the past when the climate was wetter or collected very slowly – so, if not sparingly used, are short-lived. Also, groundwater can be contaminated with agrochemical contaminants leached from landfill sites, or escaped from damaged storage tanks or pipelines, or sewage; decontamination may be very difficult and slow, if not virtually impossible. This is because groundwater often has long residence times – i.e. doesn't flush away fast – and there is less exposure to air and no ultraviolet light (sunlight) and fewer micro-organisms to attack chemical or biological contaminants. If overused, potentially sustainable groundwater resources can also be damaged or permanently destroyed through seawater or saline groundwater intrusion, or because the aquifer collapses when pumped dry.

Hopes that desalination may provide a widely used alternative water supply are unlikely to be fulfilled with currently available technology in developing countries, other than for affluent cities, areas where there is lucrative tourism or wealth generated by commerce, oil or other minerals – the costs are too high. Water supplies must be found by better management of what is already available as surface flows and groundwater, and perhaps by developing salt-tolerant crops and saline agriculture methods (see Box 3.1).

Attention is often directed toward the perceived need to install more irrigation; the task of supply dominates and the equally important co-requisite – soil drainage – often gets neglected. Consequently, irrigated soils frequently become waterlogged, which is bad enough in itself, but can also lead to chemical degradation or the concentration of salts and sodic (alkaline) compounds that contaminate the topsoil. These impacts can be difficult and costly to remedy and deplete the world's soil resources.

Box 3.1

Water: some facts and figures

- Developed country water consumption per capita (drinking, agriculture, industry and hygiene) ranges up to 3,000 litres per day (the typical amount for midwest USA) or more. The EU has suggested 1,369 litres per person per day is inadequate for a developed country, and the UN has argued 2,740 litres per person per day is a minimum. The *UN World Water Development Report 2003* suggested a World Health Organisation (WHO) goal of 20 litres per person per day of reasonable quality water, from a source within 1.0 km. Clearly, 'shortage' and 'stress' are not well defined. A poor family in a developing country will probably use less than 15 litres per day, from a polluted source.
- Domestic water distribution systems in many countries leak and waste a great deal – in most developing countries losses are more than 50 per cent, and some lose more than 80 per cent.
- Worldwide at least 1.2 billion people did not have access to safe domestic supplies in 2002. A large proportion of illness, perhaps over 80 per cent, in developing countries are water-related.
- It has been claimed that approximately 40 per cent of the world population in some 80 countries were experiencing water stress in 1998; of these nations, around 30 were suffering severe shortage. Lomborg (2001: 19–21) argued that this is based on unreliable data and there were far fewer facing shortage. In 1992, 26 countries had water scarcity; 11 were in Africa and 9 of the 14 Middle Eastern states were in a similar position.
- Roughly 70% of water supplies worldwide go to irrigate food crops, and there is great wastage doing it. World per capita grain production fell by roughly 1 per cent per year between 1984 and 1992 – it is likely that this reflects the increasing difficulty and rising costs of installing irrigation, which can often exceed US$ 20,000 per hectare.
- In the 1950s only a few developing country cities had water supply problems, now at least 26 have serious difficulties and shortages affect more than 300 million.
- Domestic supplies are commonly contaminated by one or more of the following: human waste, livestock waste, agrochemical or industrial pollution and sediment, resulting from soil erosion. The provision of effective basic sanitation, whether water-based or non-water based disposal, is vital – around 2.4 billion people did not have access to it in 2001.
- Some countries rely a great deal on hydroelectricity for energy – for example, India obtains around 22 per cent of all electricity from it.
- The problem is not just water shortages: excess water – floods – may overwhelm sewage disposal systems and contaminate water distribution systems.

Source: Compiled by the author.

Growing competition for water

For much of human history people have been able to move and settle where water and soil resources were good enough to support them; also, populations were small enough to limit overexploitation and pollution. That is no longer the case: people are much less mobile and growing human populations, rising per capita demand for water and increasing pollution have led to concern that there is, or soon will be, a water crisis (Postel, 1992; Gleick, 1993). While concern is important, adopting a crisis focus is probably mistaken and a more guarded approach is needed.

In the last decade large companies have been taking over many water supplies in developed and developing countries – some forecast that water promises to be as profitable in the twenty-first century as oil is now. While this means that what was once a free resource or a state resource is becoming privatised, this may not be all bad – because business may be better able to fund development and conservation, and to recover costs (Barlow and Clarke, 2002). Because water is so fundamentally important, it deserves priority attention from resource planners and lawyers if cartels and exploitative control by business are to be avoided.

Some current estimates suggest that by 2025 around 2 billion people will live in areas with water scarcity. Postel (1992: 28) defined 'water scarce' as anywhere with less than 1,000 m³ per person per day – effectively most of the South. Currently water managers tend to see nature and human usage as competing and at best seeking simple division of supplies between the two. That will not continue to work – human demands and nature have to be carefully integrated. As Hunt (2004) stressed, it is crucial to manage water in ways that maintain the water cycle and the ecosystems that support it as well as meet human needs, although practical ways of doing this are yet to be fully worked out. Adams (1992: 16) also stressed the desirability of '... a new approach to development, embracing integrated natural resource management with realistic socio-economic goals'. Integrated watershed management or river basin management offers possible ways forward – using biogeophysical units, with adequate overall control and powers to effectively co-ordinate the management of water, environment and human activities (see Box 3.2). Within such units water can be used as an integrative theme with its benefits distributed in a comprehensively planned way across the whole basin or watershed. Control of an entire drainage system gives administrators the opportunity to deal effectively with soil erosion, pollution, flooding and so on. Unfortunately, these approaches have frequently failed to yield their promise, but there is still growing interest in integrated or comprehensive river basin development.

Water stress, water shortage or scarcity and drought are not precise terms; consequently, comparison of data is often difficult. Recently there have been proposals for a water poverty index to give a practical and integrated assessment of scarcity (Sullivan, 2002). Societies respond to insufficient water in different ways; some innovate and reduce demand or find alternative supplies, others may abandon settlements or cease activities. There are also knock-on effects of water scarcity: poor households in developing countries commonly spend a lot of time collecting water then neglect other tasks, and in cities they may pay a large part of their incomes for supplies. Throughout the South people suffer tremendous health problems because of inadequate water for hygiene and contaminated drinking water.

The phrase 'water crisis' implies that developing countries, and perhaps the world, are currently at a point of no return or a threshold, which, if passed, will result in disaster. More objective study suggests that large proportions of the world's people are not 'in crisis' but do need better supplies. Clearly, there are huge problems with water being so grossly undervalued. The true worth of something is usually only realised when it is in short supply (Clarke, 1991; Oudshoorn, 1997; Kassas 2001; Lundqvist, 2000). There is a brighter side to this observation: if water is undervalued and wasted, then there is abundant scope to cut losses and make better use of existing resources (see Box 3.3).

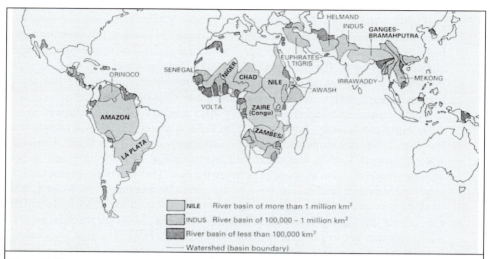

NILE River basin of more than 1 million km²

INDUS River basin of 100,000 – 1 million km²

River basin of less than 100,000 km²

——— Watershed (basin boundary)

Box 3.2

Shared river basins

Nile

The Nile is seen by some to be a potential flashpoint. It is shared by 10 states, one of which (Egypt) is very dependent on its flow for nearly 100 per cent of its water (Waterbury, 2002). In some years virtually the whole flow of the Nile is used before it reaches the sea. Egypt has a growing population and could see Nile flows reduced if the Sudan, Ethiopia or others use the river to support irrigation development. Agreements reached in 1929 to share Nile flows have broken down. The UN has responded since 1999 by backing the Nile Basin Initiative, but a sharing agreement is yet to be reached. Recently Tanzania announced it may divert large quantities of water from Lake Victoria for irrigation.

Jordan

Israel, Jordan, Lebanon and Palestine share the River Jordan. Israel uses over 80 per cent of the wastewater it produces and has little untapped groundwater and virtually no available streamflow to draw upon; its only options are to make more use of the Jordan, cut wastage, collect runoff that would otherwise evaporate or flow to waste, invest in some grand scheme to convey water from somewhere like Turkey, or desalinate salt water. There is little scope for making further use of the Jordan. Skirmishing over attempts to divert some of the flow of the Hasbani River, a tributary of the Jordan, in Lebanon preceded the Arab–Israeli War in 1967.

Euphrates and Tigris

Turkey might develop supplies and become involved in arguments with downstream states: Syria and Iraq.

Ganges and Indus

There have been strained efforts to reach satisfactory agreement between India, Bangladesh and Nepal concerning the Ganges, and India and Pakistan over the Indus, since the 1970s. There is also potential for disagreements between India and Nepal over the Ganges. The ideal of co-operative basin-wide management seems unlikely in the foreseeable future; nevertheless, there is some negotiation and agreement.

There has been a little more success sharing the Indus, although not full joint-management, between India and Pakistan.

Mekong

Six states share the Mekong: the upper basin states Myanmar (Burma) and China have so far been only marginally involved in co-operation. In 1995 Cambodia, Laos, Thailand and Vietnam (the lower basin states) signed an Agreement on Co-operation for the Sustainable Development of the Mekong River Basin. In the last few years plans for integrated river basin development for the lower Mekong have been actively pursued. Currently there are a number of developments being implemented, and a stated interest in sustainable development.

The Mekong Basin Commission is the intergovernmental body which co-ordinates the Mekong Basin states; however, China is not a member, and has been developing large dams in the upper basin which seem to be significantly disrupting flows. Estimates suggest as much as 50 per cent of dry season flow is lost, affecting fisheries and floodlands in at least four countries.

Source: Compiled by the author.

Box 3.3

Meeting water demands

Opportunities to reduce wastage of water

*   Improve irrigation and domestic distribution systems to reduce leakage and evaporative losses.
*   Supply 'on-demand' water not in large quantities or a manner which encourages wastage: one way is to use metering and realistic pricing.
*   Reduce demand – e.g. more efficient irrigation, less wasteful washing machines, smaller toilet cisterns, low moisture demand gardens; re-use of 'grey water' (non-sewage waste). There is considerable potential in agriculture to adopt approaches which conserve and better use precipitation and irrigation, which wastes less.
*   Cautious water transfers from surplus areas to shortage areas.
*   Cautious storage in dams, cisterns and reservoirs.

Opportunities to increase availability of water

*   Carefully exploit unused surface water and groundwater where it is safe to do so.
*   Treat wastewater to allow use.
*   Grow salt-tolerant crops: currently roughly 20 per cent of irrigated land is seriously affected by salts.
*   Desalination – *if* cheap energy is available or low-energy methods can be developed.

Source: Compiled by the author.

Water management

Water management should not be confused with environmental management; it is focused on engineering, economic and perhaps social goals, but not much on environmental issues. Water management is commonly poorly integrated with other activities, and those involved tend to have a short-term outlook, perhaps up to 10 years ahead, and a project focus. One group of water managers working on a dam may have little to do with others developing irrigation downstream, or with soil erosion control bodies and others. Water managers may also see themselves as serving shareholders, citizens of a city, and generally respond to project challenges without looking much further. During the last 30 years or so bodies like the IWMI and the US Corps of Engineers have been working hard to improve water management and broaden their outlook.

Some developing countries' drainage and irrigation departments were set up decades ago and have established a pecking order, which they head, and their communications with agriculture, health, environmental and other bodies may be limited and strained. This is a problem because water is involved in so many things that it must be managed in a way that is sensitive, multidisciplinary, integrated and comprehensive. Environmental managers can assist at a strategic level, advising on river systems management needs and ensuring issues are not overlooked. Central planning in the Soviet Union failed to have such a strategic multidisciplinary and environmentally aware overview; individual specialists warned of possible disasters but decision makers did not heed their advice – the result was, all too obvious with hindsight, that diversion of water for irrigation would ruin the Aral Sea region. Without environmental management overview with effective powers to act, schemes like the Three Gorges in China, developments on the Mekong, Nile or Congo and many other projects will suffer similar fates to the Aral region. The same is true for coastal zones and islands – see later in this chapter.

The environmental manager can exercise an overview and integrate, co-ordinate and prompt developments. Environmental management encourages development to go beyond just ensuring environmental quality is not compromised, to encouraging developers to seek optimum development and environmental conditions.

To date, most attention has been directed toward making water available: the reduction of chemical and bacterial contamination has been relatively neglected, especially in developing countries. The twenty-first century will hopefully see increased attention to water supply quality in developing countries, as it has becomse clear that contamination hugely affects human well being. There are of course trade-offs between keeping down costs to support economic development, and spending on pollution control and treating domestic supplies. While water-related ill health debilitates people and takes a huge toll on labour productivity, these costs are spread out and are often overlooked. Many large water development schemes may feed water to irrigation or generate power but the benefits have not been properly evaluated and weighed against their environmental, socio-economic and health impacts.

Water rights in developing countries can be more crucial than land rights. Humans cannot do without water for drinking, washing themselves and their clothes, for cooking and growing crops. Water is also needed to support industry and, a little less crucially, for hydroelectric generation and sewage disposal. While there has been great improvement in domestic supplies during the last couple of centuries, adequate provision for all is still a long way off. At the 2002 Johannesburg Summit on Sustainable Development a target was set of halving the number without access to satisfactory domestic water supplies from around 2.2 billion in 2002 to 550 million by 2015. It will be interesting to see if such a goal attracts support – poor people are not likely to offer much profit for investors.

Unfortunately, there was not much deliberation about how this improvement of supplies would be achieved; environmental managers need to ensure that the efforts do not damage the environment. Availability of adequate clean freshwater will be a key factor in any drive to achieve sustainable development (Stikker, 1998).

Some countries tap supplies of groundwater, which are not renewed fast enough to match demand, if at all. These supplies may fail suddenly and suffer aquifer changes, which prevent future recovery. Often cited is the example of Libya's Great Man-Made River Scheme: this relies on tapping fossil (non-recharged) groundwater in the south of the country and transferring it via large-bore pipelines to the north for domestic supply and agriculture. Having cost billions of US$ the supplies could fail by the mid twenty-first century, by which point many people would have become dependent on the scheme.

Knowledge about current and potential water problems is weak, and there is a need for better awareness of reserves and degradation trends. Bodies like the IWMI are seeking to improve water resource knowledge. It is important that water managers allocate enough water for wetland, lake, estuary and riverine area conservation (wetlands are discussed in Chapter 5). Quite a few cities and some farming areas depend on groundwater and have overexploited supplies, leading to falling water tables and sometimes to subsidence or contamination of aquifers with seawater or saline groundwater. Jakarta and Mexico City are examples, and Bangladesh and the Indus Basin (India and Pakistan) have had to implement costly regional salinity control programmes.

Israel has responded to shortages by making its agriculture less demanding of water and by reusing wastewater. It probably leads the world in irrigation efficiency, and has developed low-loss irrigation methods, which are being adopted elsewhere. Various dripper systems are far more efficient than established surface application or wide-broadcast sprays, and can offer water savings of 20 per cent or more. But efficient systems are currently relatively costly to install and maintain, and are thus more suitable for high-value crops rather than small farmer food production. Postel (1992: 104) estimated less than 0.5 per cent of the world's irrigation relied on drip techniques, so there is potential for great water savings if costs fall and maintenance becomes easier. Already, simple oil-drum reservoir dripper systems have been developed for small farmers, and there have been promising results with surge irrigation, which is relatively low-tech and easy to adopt and less wasteful than established large-scale gravity irrigation techniques. There is also great potential for soil and water conservation techniques like mulching, stone-lines, terraces, microcatchments and mist collection; especially if these are based on low-cost, locally available materials and labour. Soil and water conservation also reduces soil degradation and improves groundwater recharge.

Water management demands awareness of what is happening beyond aquifers, river channels and lakes, and if possible some control over such things. For example, even slight changes in vegetation cover can affect runoff and water quality, and can alter groundwater levels and cause salinisation of soils. Poor land management means soil and agrochemicals contaminate streams, groundwater and lakes. However, in some developing countries water authorities do not liaise with departments of agriculture, industrial developers, city supply bodies and so on. And, as mentioned earlier, there may be rivalry and secrecy, making integrated management difficult.

Acid deposition presents a growing threat for developing countries as industrialisation spreads (Rodhe and Herrera, 1988). Acid pollutants may accumulate on tree foliage, in snow cover, attached to clays in the soil or in the bottom mud of waterbodies, and later be suddenly released by a spring thaw, ploughing of farmland, dredging of waterways, autumn leaf-fall and possibly global warming, leading to a concentrated 'flush' into streams and groundwater. Accumulated pollutants may remain 'locked up' in soil or sediments for decades after the source has ceased, and then escape as a consequence of environmental changes or disturbances. Such sudden pollution threats have been called 'biological or chemical timebombs'. The disaster happens once a crucial threshold is passed, and may be avoided or prepared for if environmental managers understand the risk, monitor conditions and ensure authorities take adequate action. Water managers must also be vigilant for bioaccumulation (bioconcentration or biomagnification) effects. For example, lake water may have 50 ppt of a pollutant, which can be concentrated by plankton that are then fed on by fish, which are in turn fed on by predatory fish – so that at the top of the food-web, levels can be as high as 40,000,000 ppt in something like turtles or fish-eating birds (Stauffer, 1998: 45).

Sewage pollution is a widespread problem in developing countries, as it can result from livestock waste, storm drainage from urban areas and normal sewage discharge. Agrochemical runoff from farmland and effluent from mining and minerals processing, sugar, oil palm, rubber, aquaculture, alcohol production and other agro-industrial activities also cause difficulties (see Box 3.4). Where problems are caused by agrochemicals it may be possible to adopt less dangerous compounds, or to apply less in more focused ways; even better, timing of applications or controlled-release formulations can cut pollution. Also, it should be possible to make more use of alternatives like biological control – using predators, fungi, or micro-organisms which control weeds and insect pests or use genetic 'engineering' to discourage or disable pests, and produce crops that resist insect and weed attack and need less fertiliser.

Sewage and livestock waste can be utilised for methane production, sent to composting plants, or dried and burnt as fuel for district heat and electricity generators. But the waste must be of a suitable quality, the supply adequate and the investment available to install biogas digesters, composting facilities or furnaces and generators. In peri-urban areas of developing countries sewage might be effectively used for agriculture, if there are careful controls. This is not a new idea,

---

Box 3.4

Responses to water-borne pollution

Pre-1840s in developed countries/currently in most developing countries: simple short distance discharge/dumping.

1840s to present: pipe effluent away to river/lake/sea and hope it dilutes enough to cause no obvious problems.

1930s to present in developed countries/currently in many developing countries: 'end-of-pipe' solutions. An attempt is made to treat the pollutant to make effluent less harmful with primary (very basic) sewage works, treatment lagoons for industrial plants, etc. Seldom is the treatment effective enough; the processing may break down or be overwhelmed by heavy rainstorms, so pollution escapes; the processing can be a nuisance to neighbouring areas; processing generates concentrates (sewage sludge) which pose disposal problems. Also, this approach does not address 'non-point' sources well – i.e. pollution from dispersed sources is difficult to collect and process.

Late 1980s to present in developed countries/here and there in developing countries: prevention/ avoidance; restoration of degraded environments; waste-utilisation. These overcome many of the aforementioned problems. Also, some pollution is channelled to where it is a useful raw material: e.g. brick-making with waste silt; biogas generation; sewage sludge composting; waste-water irrigation. Adoption of approaches which satisfy the precautionary principle where possible.

1990s on in developed countries spreading in developing countries: more interest in appropriate, low-cost solutions such as separation of highly-contaminated wastes and slightly polluted 'grey water', with the latter used for various non-drinking/no health-risk useful purposes; renewed interest in use of aquatic plant-filled lagoons to process wastes, and in return of sewage to the land (with or without composting).

Note: Historically there is almost always a lag between pollution and a response. In developing countries it may be possible to 'leapfrog' – i.e. to adopt new technology, often without incurring development costs and teething problems, which developed countries have to wait to update to.

Source: Compiled by the author and various sources, including Stauffer (1998).

---

and it also assists with disposal (see Chapter 7). There are of course disease risks and the danger of contamination of produce with heavy metals and other harmful compounds; however, sewage could be chlorinated cheaply and harmful wastewater from storm drains or industrial areas could be diverted. If used for crops like trees, fuelwood, fodder or amenity area irrigation, rather than food crops, the disease, heavy metals, and other contaminant risks would be less important. If authorities do not undertake improvement and distribution of sewage, there is a risk that urban and peri-urban peasant farmers will make unsupervised use of contaminated water.

- Digging drainage and irrigation channels, roadside ditches and canals, the movement of tourists and manufactured goods, and accidental or deliberate introductions, can all alter the distribution of plants and animals. Some organisms spread and markedly affect water resources and human health; for example:

- Various burrowing crab species have managed to spread to new locations in the ballast water of ships or on fishing gear, and cause serious damage to native species, banks and channels.
- A number of aquarium and aquaculture escapees seriously compete with native aquatic species. Careless fisheries introductions and aquaculture escapes also cause problems.
- Ostracods – aquatic animals, usually smaller than a few millimetres in diameter – can harbour the organism responsible for cholera and have been spread in ship ballast to infect new areas. Serious outbreaks took place during the 1970s in Peru and other parts of South America, having originated from South Asia. Filtering and/or disinfecting ballast water can reduce these problems at a reasonable cost.
- The zebra mussel (*Dreisseena polymorpha*) and the water hyacinth (*Eichornia crassipes*) often colonise and choke waterbodies, channels and pipes, raising maintenance costs and causing other problems throughout the tropics and sub-tropics.

Some developing countries have responded to water supply problems by importing cheap foreign grain and neglecting their own agriculture (Yang and Zehnder, 2002). This is because the costs are lower and because tastes, especially those of more numerous city-folk, have shifted from traditional crops to imported foodstuffs, particularly wheat and maize. If this shift from traditional food production to cheap grain import expands there could be a number of impacts. It would probably affect global food reserves and more people will depend on fewer producers, which means increased risk. In the grain-importing countries, rural employment will probably deteriorate, reinforcing the drift of people to cities and into relocation to developed countries; and it will probably increase environmental degradation in the countryside as traditional land use gets neglected. So addressing water supply problems can have many indirect benefits.

Developing rivers

The usual response to water shortage in developed and developing countries has been to seek a 'technical fix' – often large-scale engineering aimed at storing riverflow or transferring flows from water-rich to water-poor areas. Barrages are not as much of a barrier to flow or riverine species as dams. Barrages divert some flow from a river, so there is no large reservoir; however, they do not store water to meet demand in dry periods, and give limited 'head' for electricity generation. Where water storage is needed, and where power generation is wanted from a river with fluctuating flows, a dam is the usual solution.

Large dams have been constructed on virtually all major rivers worldwide since the 1920s. Between 1950 and 2004 over 35,000 large dams have been constructed, mostly in developing countries, and more than 19,000 in China alone. About 400,000 km² have been inundated by dams globally and it has been claimed that between 40 and 80 million people have been displaced since 1960 (http://www.gfbv.de/gfbv_e/uno/geneva03/item_7_en.htm, accessed February 2004).

After the Organisation of Petroleum-Exporting Countries (OPEC) oil price rise of 1973 there was incentive for debt-ridden developing countries to cut petroleum imports; hydroelectricity seemed to offer this, and give other benefits like irrigation supply. The construction of large dams reached its peak during the 1970s (over 5,000 were built in that decade) and there has been a slower rate of increase since, especially after the early 1990s, partly because of strong criticism (McCully, 2001). There were signs of renewed interest at the 2003 World Water Forum where calls were made for more dams (Pearce, 2003). There are currently plans for a huge hydroelectric scheme on the Congo, the first dam to be sited at Inga Falls, to serve a proposed pan-Africa electricity grid. And other schemes are proposed or under construction elsewhere.

Dams and barrages can cause difficulties by altering the flow of water and sediment and restricting the movements of fish and other organisms. Dams also form large reservoirs, which often pose difficulties. At first those involved in dam construction had to make do with inadequate knowledge about tropical riverine ecosystems and the impacts of regulation and impoundment; as time passed experience should have helped reduce unwanted problems. Unfortunately, similar mistakes have been repeated time and again. It is striking how slowly engineering and dam management has become more appropriate and flexible. Some of these problems result from insufficient attention to environmental and social management; others are inherent in large-scale modification of river flows. Some seem to result from what Newson (1992: xx) calls a 'point-problem' outlook; i.e., the developers do not want to look upstream or downstream at the whole issue. Consequently, river development is overshadowed by inflexible and insensitive 'solutions', which for political, economic and cultural reasons continue to be implemented (Adams, 1992: 170; McCully, 2001).

The World Commission on Large Dams (WCD) was established in 1998 by pro- and anti-dam interests keen to debate and review developments and form an independent body to set guidelines and assess alternatives. The WCD is due to terminate in 2005 after publishing its final report (WCD, 2004). An ongoing body interested in improving large dam planning and management is the UN Environment Programme Dams and Development Project. One dam problem does seem to be getting more attention – only quite recently has serious consideration been given to releasing artificial floods from dams to try to maintain downstream environmental conditions. The reluctance to do so is probably because managers want to retain water for power generation or irrigation – a water management focus rather than a wider environmental management focus. Similarly, a number of countries have spent a great deal on flood protection through construction of levees – raised banks alongside a river. This seldom gives adequate flood protection, can shift flooding to less-protected areas, and prevents flood silt from fertilising riverside lands. Worst of all, if a levee is breached flooding tends to persist because drainage is hindered and the people may have had a false sense of security and were little prepared. Again, it is interesting to see approaches being used long after they have been proven harmful; the late-nineteenth century engineer who virtually established large dams – William Willcox – was very aware of the need to let silt and floodwater pass downstream from dams and of the problems caused by artificial levees.

Riverine ecosystems are one of those most affected by humans; people often live close to rivers, dispose of their waste in them and rely on them for drinking water, irrigation, transport and power. There has been some success in countering human waste pollution and some industrial effluents in developed countries in recent years, but this still presents a major challenge for environmental managers in developing countries (see Chapter 7). Worldwide, the range of pollutants entering river systems and groundwaters has greatly expanded, to include: nitrates, phosphates, heavy metals, polychlorinatedbiphenols (PCBs), dioxins, endocrine disrupters, oestrogens ('gender benders') and many others. Some of these can suppress the immune systems of wildlife and humans or cause reproductive problems and cancer even at low levels of contamination (Malmqvist and Rundle, 2002).

Numerous grand-scale interbasin transfers have been proposed, some have failed to get beyond the drawing board and a number have been abandoned. For example, warfare in the early 1980s put a stop to the Jonglei Canal, which was to channel water past the Sudd swamps in the Sudan and increase supplies in the mid and lower Nile. Interest in large integrated basin development schemes and grand interbasin transfers is currently growing. The latter are planned for Latin America (Grand Hidrovia canal system), and China is pressing ahead with a huge (over 1,200 km) south–north water transfer scheduled for completion in 2007. This will channel flows – equivalent to about a tenth of that of the Mississippi – from the Yangtze River to Beijing. India has proposed the Indira Ghandi Canal and a national water grid intended to redistribute river flows on a continental scale. Large-scale transfers might help resolve water shortages in the Middle East if conflicts are resolved and terrorism becomes less of a threat. Developing countries should note that the North American Power and Water Alliance has demonstrated that any large-scale transfer needs to carefully consider social, economic and environmental impacts. Before embarking on huge, costly and probably inflexible river transfer schemes it should be asked whether a better approach would be to control water wastage and discourage further growth in the water scarce areas, rather than channel in supplies which may generate problems and perhaps lead to greater water demand in future years.

Most of the world's larger rivers are shared by more than one country. If tension grows over a shared river there are two possible ways forward (Klare, 2001; Ohlsson, 1995):

1   Consultation, agreement and co-operation
2   Power politics ('hydropolitics') and possibly conflict.

A somewhat sensationalist literature has been warning of the risk of 'water wars' since the 1970s. However, so far, interstate disputes have almost never led to conflict. As demand for water increases and more supplies are contaminated, the risk of conflict could rise (Ashton, 2002). Swain (2001) noted that agreements on sharing rivers may be possible while say 80 per cent of total flows run to the sea, but feared if future demands rise, and if environmental change and pollution cut supplies, agreements reached may not hold and new ones would be less likely.

The goals of shared river management should be: peaceful agreement, sustainable management, social equability and opportunities for integrated and comprehensive development and maintenance of environmental quality. Much of the negotiation and agreement so far have just focused on quantity of flows, less on quality, yet the problem of contamination is growing – and few address the other goals just listed. A Convention on the Protection of Transboundary Watercourses and International Lakes was signed in 1996, but is only a start to resolving water quality problems.

Laws relating to sharing of river flows and groundwater are not very well developed. The UN Convention on the Law of the Non-Navigational Use of International Waters was adopted by the UN General Assembly in 1997 and provides a framework for negotiations between countries. This defines the obligation not to cause harm to another and the right to reasonable and equitable use by riparian nations – but it is not law.

Large lakes

Large natural lakes have suffered in the last few decades in developed and developing countries as a consequence of pollution with sewage, industrial effluent and agrochemicals; problems have also been caused by the introduction of alien species of plants and animals; the diversion of inflowing rivers, often for irrigation supply; over-exploitation of fish, game and other animals; and disturbance related to tourism. Some lakes in developing countries are of great age, and because they lie at low latitude their biota are diverse and have escaped the worst impacts of Quaternary environmental changes, but their rare endemic species are now vulnerable (Beeton, 2002). The former Soviet Union has the world's largest and deepest freshwater lake (Baikal), which has suffered considerably from industrial pollution and paper-pulp production effluent.

Agricultural or industrial activities cause much lake pollution; for example, the production of cut-flowers for export around some of the African Great Lakes (see Box 3. 5). Expanding tourism and associated sewage has despoiled many lakes, like those in Kashmir. Aquaculture can cause serious impacts through waste effluent, escapes of competitive species, introduced diseases and use of chemicals, which may affect lake organisms. There should be efforts to encourage and support the development of less-polluting and sustainable aquaculture approaches, and vigilance for damaging activities. Lake management needs overall co-ordination, to integrate all the relevant fields, and stakeholders, and should preferably be administered through a single lake development authority with adequate enforcement powers – this may need to be multinational.

Irrigation, runoff cultivation and improved rain-fed agriculture

Huge effort and expenditure has been directed at adapting land to fit crops; little, until relatively recently, has been spent on adapting crops to fit the environment. Social scientists would also add that in the same period the norm has been to make people fit innovations, rather than innovate in ways that suit their needs and offer

Box 3.5

Biodiversity loss in the African Great Lakes

Lake Victoria in East Africa is shared by Kenya, Tanzania and Uganda, and is the shallowest of the African Great Lakes; it may therefore show signs of pollution before others and provide a warning. Over the last two decades more than 200 species of endemic chichlid fish, of an original population of about 500 species, have become extinct. The blame has been placed on the introduction of the Nile perch, one of several alien species introduced in the 1950s to compensate for overfishing. However, recent studies show that it is more likely that pollution has been the main cause of the species losses. As a consequence of the deforestation of surrounding land, modernisation of farming and increased sewage pollution, the Lake has become eutrophic, i.e. the water is overrich in nutrients. This has led to increased algae and phytoplankton, which make the water murky. Chichlids rely on visual stimuli when breeding and without clear water there has been a failure of some to breed; others have interbred, which also threatens biodiversity.

The damage has now been done; reducing pollution will not restore the original biodiversity. Hopefully, environmental managers will note the situation and act to reduce similar problems in Lake Malawi and Lake Tanganyika.

Source: De Weerdt (2004).

sustainable results. The usual strategy has been to develop large-scale commercial irrigation which is wasteful of water, difficult to sustain and liable to contaminate rivers with agrochemical and salt-contaminated return flows. However, there are alternatives, notably runoff cultivation, improved rain-fed agriculture and more support for small farmers. So far these alternatives have had much less investment, although they may be the only practical solution for many regions. The distinction between irrigated, runoff and rain-fed agriculture is not always clear-cut; indeed, there is overlap. Also, improved moisture use, sustainable agriculture, and soil conservation are generally closely interrelated because wasted water tends to flow and erode the land and take soil and nutrient-rich debris with it.

There are many ways of improving rain-fed agriculture (i.e. reliant upon precipitation and not irrigation), including: developing shorter growing season crops, the introduction of tractor ploughing, drought-resistant crops, fallowing, careful use of fertiliser and green manure, rangeland forage improvements, stall-feeding livestock, and other innovations (Barrow, 1987). These do not demand much water, compared with usual irrigation development, but can improve security of harvest, boost yields and through soil conservation allow sustainable production. There are also opportunities for developing crops which can make use of saline water or salty soil, or which fix atmospheric nitrogen and effectively provide fertiliser for themselves. The established methods of irrigation – mainly large-scale canal-supplied, gravity-fed schemes – are increasingly difficult to expand, because most of the more suitable sites have already been developed. Also, water is less easy to come by, and there are significant cost increases – installing such schemes today can cost a lot. Sustaining large-scale irrigation is also a challenge, and it is not unusual for a project to fail to repay its investment costs before falling into disrepair. The hectarage of irrigated land per person worldwide has shifted from marked increases during the 1960–1990s to present day decline (Postel, 1992: 50–1).

Large-scale irrigation not only wastes water and damages the environment, it can also make people ill – especially by increasing the transmission of malaria and shistosomiasis. Roughly two-thirds of the world's irrigation is in need of repair, but it is still common for schemes to be abandoned and new ones started elsewhere. Huge sums of money have been spent on large-scale commercial irrigation and some of the problems encountered are similar to those of large dams: inflexibility, insensitive engineering 'solutions', wishful thinking by outsiders, reluctance to work with local people, and a failure to learn lessons in spite of considerable hindsight experience (Adams, 1992). Some at least of these problems would be resolved by more integrative management. There is one large irrigation scheme in Peru where small traditional plots were converted to large-scale carefully graded fields. If environmental studies had been made and local knowledge had been gathered it would have been clear that the region suffered flash flooding and frequent earth movements – small fields were adaptable and ideally suited, large level fields were vulnerable. Consequently the development failed at huge cost.

One way people can obtain drinking water, grow crops and water their livestock is by means of a range of rainfall harvesting techniques – runoff agriculture. These are part runoff control and part soil conservation methods. Archaeology has helped prompt interest; studies in the Negev Desert of Israel, north Africa, the Andes, south-western USA and Latin America have yielded promising strategies. Parts of India, Sri Lanka, Africa and the Middle East have long traditions of effective soil and water conservation techniques, which make use of local labour and indigenous materials, and these are still being practised. In the past, peoples in a number of regions used techniques which modern agriculture would be hard put to match, and in some cases fed hundreds of thousands in harsh environments.

It has become apparent that some of these strategies could be appropriate for modern small farmers and larger producers in developing and developed countries. Such approaches are inexpensive and do not condemn users to dependency on outsiders; techniques can often be improved by modern inputs, and they may demand little that is not available locally to poor farmers. Some of the techniques are potentially excellent sustainable agriculture strategies. In the Negev experimental farms copying 1,500-year-old methods discovered by archaeologists have shown that crops can be sustained even where there is less than 100 mm of precipitation a year. The strategy is to use runoff collection from catchments larger than the cropped plots, or by digging microcatchments which concentrate moisture and soil around a single tree or patch of crops. The latter can be used on virtually any gradient if soil and other conditions are suitable, and if there is a little precipitation during most of the growing season and relatively low evaporation (Evanari et al., 1982; Reij et al., 1996; Barrow, 1999b).

Rainfall harvesting has great potential for providing more food, making cropping more secure and sustaining land use in poor and remote areas. Such locations are unlikely to get much support from commercial organisations so accessible, low-input techniques are crucial. These soil and water conservation strategies can usually gradually improve the land, which makes sustainable development more likely. But the people involved must be motivated and have access to sufficient social capital to adopt and sustain the approaches – archaeology suggests that social change and environmental upsets can wreck even the best-established runoff harvesting systems, so social institution building is a crucial element.

There have been various successful runoff agriculture developments already. One has been the spread of stone-lines in West Africa. These are simply rows of cobbles placed along the contour to slow runoff, retain soil and moisture, and so improve cropping and reduce land degradation. Their adoption is taking place with limited outside aid because they have gained popularity, can be maintained locally, improve security of harvest and offer better yields *with little risk or excessive labour input incurred by the adopter* (Barrow, 1999b).

Far less has been spent developing small-scale irrigation or rain-fed and runoff cultivation than has been expended on large-scale irrigation. It is possible money would have been better spent on such alternatives; but it would be rash to assume these would certainly have fared better; small is not necessarily beautiful and social factors can be critical.

Better management of water resources

Water resource management needs to be comprehensive, integrative and environmentally sensitive and socially appropriate. Water resources managers have discouraged projects with too narrow a focus on economic and engineering goals. Growing competition for finite supplies means better ways of co-operating and sharing are needed. Maintaining the quality of supplies also demands a more strategic overview. Getting such improvements will not be easy and will require environmental management inputs (see Box 3.6).

Coastal zones and islands

Coastal zones contain more than 50 per cent of the world's population – some sources suggest 75 per cent – and many of the biggest cities, a large part of all industry, and much other human activities take place in these regions. Coastal zones are thus subject to diverse and intensive use; they comprise roughly 10 per cent of the Earth's surface, and are vulnerable to storm damage, tsunamis, sea-level rise and pollution from the sea. There is no precise definition, but they are usually accepted to be the belt of land bordering the sea inland to about 20 km. The intense human activity there often results in serious environmental challenges.

The coastal zone includes estuaries, salt marshes, mangrove swamps, coral reefs, near-shore islands, mudflats and so on. Not surprisingly, this complexity has led to increasing use of integrated, comprehensive or holistic planning and management approaches – pursued through a coastal zone management strategy and a coastal zone management authority (Brown et al., 2002; Mermet, 2002; Mohammed, 2002; Cicin-Sain and Knecht, 1998; Clark, 1996). There are similarities to integrated river basin management in that development is taking place in a discrete, non-ephemeral, well-defined biogeophysical region which can be studied using systems approaches or just treated as a regional development unit. In the USA and Europe, coastal zone

---

Box 3.6

What is inhibiting environmentally sound water management?

- Debt/lack of funds;
- Variable conditions from project to project (tropical biological diversity especially) so difficult to generalise and predict in advance;
- Foreign expertise is keen to get 'results' and to move on as fast as possible, leading to hurried work and lack of feeling for local conditions: accreditation by an international professional body may help resolve this;
- Shortage of local people with skills;
- Poor cost recovery (water too cheap);
- Planning and management do not consult 'beneficiaries' enough;
- Poor baseline data;
- Institutional weaknesses (e.g. departments involved may not communicate);
- Funding body doesn't liaise well;
- Politics and history hinder things;
- Useful research neglected because it is unfashionable or offers poor career returns;
- Ministries, companies and experts do not exchange information;
- Expedience and special interest groups dominate development;
- Experts do not 'get boots muddy': i.e. fail to go into field enough and get first-hand experience;
- Centralised decision-making: local managers have to consult headquarters before action can be taken;
- Skilled managers are keen to leave and return to cities where there are better living conditions or to move to new projects with career opportunities.

Source: Compiled by the author from sources including Thana and Biswas (1990).

---

management is well-established, and a number of developing countries have integrated coastal zone management policies (Mokhtar and Ghani Aziz, 2003). Coastal zone management has its own specialist journals, like *Ocean and Coastal Management* and *Coastal Zone Management*.

Offshore islands are essentially coastal zones with limited hinterland. They are also more isolated than coastal zones. Islands, especially those that are small and remote, can be very vulnerable environments. The most isolated tend to be oceanic islands, and a high proportion of their biota may be endemic – specific only to that one locality. Endemic species are often poorly able to compete with alien organisms, so they easily become extinct.

Environmental management of coastal zones and islands needs to co-ordinate diverse, sometimes conflicting, activities, which may include industry, tourism, aquaculture and fisheries, coastal erosion control, power generation, dredging of sand and gravel, seaweed collection, conservation, waste disposal, and many others. Coastal zone and island ecosystems often share terrestrial and aquatic features. Some, like estuaries and mangrove forests, depend on terrestrial ecosystems to supply freshwater, nutrients and so on. Yet, like coral reefs, lagoons, seagrass meadows and saltmarshes, they may be vulnerable to pollutants washed from the land.

The world's coastal zones and low-lying islands will be much affected if projected global sea-level rise takes place (Ince, 1991). The scale and seriousness of this threat means that plans should be underway now to try to cope with the problems; in some situations abandonment and retreat may be the best option, leaving the land to become mangroves, marshland or beach which could be used for biodiversity conservation, aquaculture, grazing, tourism and storm protection. Other areas may decide to use barriers and banks to try and hold back the sea. People may be forced to relocate from some existing coastal zones and islands, and there could be huge numbers of these eco-refugees. Many Pacific atolls face a bleak future if there is even slight sea-level rise. They are aware of this and have started to lobby the international community through the Association of Small Island States (AOSIS). Islands have already suffered through outsiders: some have been seriously degraded in colonial times, resulting in the loss of forest, topsoil or nuclear contamination. A few were taken for military bases or prisons and the indigenous people were exiled – examples include Diego Garcia, and the Andaman Islands. And various islanders are currently seeking compensation, improved human rights or a chance to return and resettle.

A number of countries have invested in aquaculture, largely aimed at producing luxury export products to win foreign exchange. Private sector tropical prawn and shrimp production is particularly lucrative and in many countries has caused severe environmental problems and socio-economic impacts on local people. It is difficult to control because owners make money and become powerful, and they generate foreign exchange, so government may hesitate to restrict them. Tourism management is often a significant issue in coastal zones, and can result in large numbers of visitors even in quite remote areas.

Coastal zone management faces particular challenges where there are 'enclosed' or sheltered seas, lagoons, coral reefs or mangrove swamps – all especially sensitive environments which are likely to be less quickly flushed clean by the sea than open coasts. These support vulnerable biota and are generally badly damaged if they are polluted. Reefs are also problematic. Worldwide, reefs have suffered damage from fishing nets, coral collectors and pollution. In some cases the damage is directly due to pollution by agrochemicals or sediment washed from the land. Sometimes it is more indirect: widespread damage has been caused in the Western Pacific and the Indian Ocean by the crown-of-thorns starfish (*Acanthaster planci*). Whether the creatures initiate the damage or are favoured by reef decline is not clear, although they appear to have boom growth every 15 or so years, and these infestations seem correlated with increased phytoplankton caused by nutrient runoff due to agrochemicals and cultivation. There is considerable concern over coral bleaching, which looks likely to be caused by rising sea temperatures. The Great Barrier Reef off Australia and a number of remote Pacific atolls are affected. The situation is being monitored by the Global Coral Reef Alliance based in New York. Reef bleaching damage results in serious loss of biodiversity and exposes shorelines to storm damage.

Some coastlines and islands at risk from tsunamis have early-warning systems; much of the Pacific is covered by a US-operated service, which can give a few hours' alert, and Japan has a regional network giving several minutes' alert. Coastal

Figure 3.2 *A crowded coastal zone. Urban growth since the late 1970s has all but covered available building plots between the sea and hills, forcing Georgetown, on Penang Island, Malaysia, to grow upward and out into the sea. View from 65-storey KOMTAR Centre.*

Source: Author, 2002.

areas and islands of the Atlantic have little protection at present (Barrow, 2003: 179–81). Warning of hurricane, typhoons and other extreme weather events has improved a great deal and most places get reasonable warning. Island wildlife is vulnerable. Many species have been lost through disturbance, vegetation clearance and hunting, and as a consequence of alien species arriving as stowaways or deliberate introductions. Rats and cats are a widespread cause of losses, but many other creatures threaten biodiversity, ruin crops, cause a nuisance and spread diseases. Ongoing monitoring is vital if problem species are not to become too well-established to dislodge.

With little space development pressures are soon felt – 5,000 people on a small island can exert huge demands (see Figure 3.2). Lacking economies of scale, and having to import many things often from great distances, it may be difficult to find funds for environmental management. Employment opportunities are likely to be desperately needed and consequently discouraging environmentally degrading activities can be a challenge. People with education and initiative may migrate away, leaving a conservative and often elderly population which may have little energy and money for environmental care, or interest in fighting damaging activities. On the other hand, islands and coastal zones are a discrete and often manageable-sized unit for the environmental manager to work with; in some cases, they are affluent tax havens, oil states, rich fishing economies, or are attractive enough to lure tourists and rich retirees. So, there are islands and some coastal zones, like the Persian Gulf, with no shortage of funds. When there are funds the environmental manager is likely to have powers and support to achieve a lot – and here and there islands and coastal zones are at the forefront of sustainable development and environmental management.

The assumption should not be made that developing countries accept that environmental management is 'for the public good' or that people should have a say in what happens to anything other than their local surroundings. As already argued, environmental management is being disseminated from Western 'liberal democracies' and it must adapt to local socio-political conditions. Turnbull (2004) provided a study of the complexities of applying environmental management in the Fiji Islands, where indigenous class and power structures may not readily support established environmental management approaches. Currently, the South Pacific Regional Environmental Programme (SPREP) is seeking to spread and institutionalise environmental management. SPREP is a Samoan-based body, which is promoting environmental management in 22 Pacific island nations.

Marine pollution

Around two-thirds of the Earth's surface is ocean-covered, and many developing countries have claimed territorial waters to at least 12 miles, some as far as 200 miles, from their shores so as to have jurisdiction over minerals or fisheries. Other than these claims, oceans are in large part beyond sovereignty. A growing number of pollutants have dispersed to the world's oceans and are now present at worrying background levels.

Significant problems are also caused by oil spills from bulk-carriers washing their tanks or pumping out oil-contaminated ballast water. There is also leakage from oil production sites, and catastrophic spills occur when there are accidents. Control is largely a matter of monitoring and enforcement of deterrent measures to try and prevent tank washing. Reconnaissance using aircraft and satellites can now spot oil slicks and there is a good chance the source chemical 'signature' can be identified accurately enough to prosecute offenders. Costs put this sort of monitoring and enforcement beyond most developing countries, so international funding or the polluters must cover it.

Catastrophic spills can be more or less contained with various floating devices to be skimmed-up and safely disposed of, *if* the sea conditions are favourable and *if* an organisation with equipment is close enough and someone is prepared to pay. Spills that escape such measures are often dealt with using detergent which may do more harm than good. Tropical mangrove ecosystems are especially vulnerable to oil spills; many species spawn in these ecosystems or use them as nursery areas and they are often important sites for aquaculture and fisheries (Vannucci, 2001). Once polluted, mangroves and coral reefs suffer badly and may not recover.

The sea is becoming increasingly polluted by chemicals and rubbish washed down rivers or sewers. A major problem is caused by plastic debris which float considerable distances. Plastic bags can cause serious injury to turtles and other wildlife when they are ingested. The problem at sea and on land can be reduced if plastic packaging is biodegradable, but the breakdown may be much faster in sunny and warm developing countries than in the cooler developed nations where the materials are usually developed, so the materials will have to be 'tropicalised' or consumers in warmer environments will have packaging degrading prematurely.

Some oceanic pollution is from refuse discarded from ships that are supposed to store waste and then dispose of it properly in port. Having recently smelt a luxurious cruise liner from nearly 2 km downwind – and gagged – I understand why the law may be flouted and waste gets tipped overboard. Some waste is a consequence of storm damage – plastic sheeting torn free and blown out to sea or material washed off ships during heavy weather. There have been numerous reports of considerable waste even in remote ocean locations, so there is obviously a need to tighten controls.

Concluding points

- Water resources are under stress and the situation is getting worse. These resources are undervalued, widely misunderstood and mismanaged. Water resources management approaches have been too narrow and environmental management must play a co-ordinating and integrating role.
- Water must be much better managed. Technology should not be discarded, but a more flexible and sensitive application is needed and management must have a broader overview, which includes environmental and social awareness.
- Agricultural improvements since the 1950s have been mainly based on irrigation: often insensitive and poorly managed schemes which frequently perform poorly. Rather than alter environment to suit crops it may be better to adapt crops and techniques to suit the environment – for this, soil and water conservation have potential.
- Coastal zones contain more than half the world's population and are coming under considerable and varied pressure from development.
- River basins, coastal zones and island environments are heavily settled, subject to diverse demands and facing challenges. These three biogeophysical and development units are some of the most altered and vulnerable parts of the world. They are probably best managed through comprehensive and integrative planning and management, overseen by an authority with adequate jurisdiction and powers.

Further reading

Hunt, C.E. (2004) *Thirsty Planet: strategies for sustainable water management.* Zed, London. [Readable and radical coverage of freshwater ecosystems which focuses on human usage and management. Presents alternative approaches, which should protect nature as well as supply people with water.]

McCully, P. (2001) *Silenced Rivers: the ecology and politics of large dams* (updated edn.). Zed Press, London. [An excellent text on dams and large development schemes.]

Adams, W.M. (1992) *Wasting the Rain: rivers people and planning in Africa.* Earthscan, London. [Water development in the real world. Excellent critique of inappropriate approaches.]

Reij, C., Scoones, I. and Toulmin, C. (eds.)(1996) *Sustaining the Soil:indigenous soil and water conservation in Africa.* Earthscan, London. [Fascinating coverage of indigenous soil and water conservation – water and land use should not be seen in isolation.]

Websites

International Water Management Institute – improving water and land resources management for food, livelihoods and nature: http://www.iwmi.cgiar.org/ (accessed April 2004)

International Rivers Network encourages equitable and sustainable methods of drinking water and energy supply, and flood management; via http://www.mylinkspage.com/earthsummit.html (Earth Summit Information website Section 46 provides this and other water websites) (accessed April 2004)

Clean water for poor people in developing countries from One World Action (London): http://www.oneworldaction.org (accessed October 2002)

Coverage of large dams – UNESCO: http://www.unesco.org/courier/200;World Commission on Large Dams: http://www.dams.org (accessed January 2004)

Soil and Water Conservation Society: http://www.swcs.org/t_top.htm (accessed February 2004)

International Center for Living Aquatic Resources Management (ICLARM) – aquatic resource management research; development of aquaculture and other productive management approaches: http://www.cgiar.org/iclarm (accessed April 2004)

Coastal zone management: http://www.coastalmanagement.com/ (accessed April 2004)

Bibliography on coastal zones (Island Resources Foundation) – general and Caribbean material: http://www.irf.org/irczrefs.html (accessed April 2004)

4 Agriculture, land degradation and food security

Key chapter points

- This chapter examines food production and hunger, and ways to improve yield, security and sustainability.
- Hunger is a serious problem and a threat, which could still menace even developed countries. Food production and food security need more attention and investment.
- Agricultural improvements since the 1950s have been masking worsening soil degradation and are causing agrochemical pollution and soil degradation. The Green Revolution strategies used to improve food supplies rely on techniques and inputs which hinder or prevent sustainable development, and in some areas production is declining or looks likely to do so. In the future agricultural improvement will need to take the form of a Doubly Green Revolution which boosts yields, sustains the production, causes much less environmental damage, and copes with global environmental change and pollution. It is also desirable to ensure there are innovations suitable for small farmers and herders as well as commercial producers.
- Soil degradation and desertification threaten food production and biodiversity. These signs of mismanagement deserve careful study to avoid misconceptions, which hinder control measures.
- One route to resolving some of the current challenges faced by agriculture is genetic engineering; this could allow humans to respond more effectively and rapidly to problems and opportunities. While genetically modified organisms (GMOs) might help provide sustainable and less environmentally damaging food production, they could equally run out of control and pose serious threats. Indeed, GMOs have already caused problems in a number of countries. Environmental managers must monitor GMOs to maximise benefits and to help ensure nothing goes wrong.

Technological advances in agriculture since the 1940s have probably been masking soil degradation. Increasing pollution in some regions has already started to limit cropping with established approaches. Global environmental change is likely, so agriculture should be adaptable to cope. Rising populations and demands for better standards of living will put agriculture under pressure. The strategies mainly used at present place the emphasis on yield increases, and much less on sustainability, security of harvest, equitable access to produce and employment generation. Much of today's agricultural research and investment is directed at larger-scale commercial producers in favourable locations. There is less interest in the world's small farmers and herders who often live in relatively harsh and remote environments.

In many regions small farmers and herders are suffering a breakdown of traditional livelihood patterns as a consequence of various development pressures. Agricultural improvements are needed to counter such problems and to try to control increasing poverty, land degradation, urban migration or emigration.

Land and soil degradation

Knowledge about natural soil qualities and distribution, let alone current fertility, degradation status, or vulnerability to various threats such as global warming or acid deposition, is inadequate. The United Nations Educational, Scientific and Cultural Organisation (UNESCO) and the UN Food and Agricultural Organisation (FAO) published a *Soil Map of the World* in 1974, and in 1990 conducted a Global Assessment of Soil Degradation, which provided an estimation of the extent and severity of soil erosion and decline in fertility – although at a coarse-scale. Environmental managers need better soils data than is often available; unfortunately, soils specialists are in short supply.

There are many soil managers, farmers and herders, many of whom are unable or unwilling to make a good job of it. Unless an adequate portion of the profits or sufficient labour is reinvested in land husbandry, yields will fall and soil will degrade. Failure to take care of the soil may result from poverty or population growth, insecurity of some kind, insufficient income to pay for soil upkeep, or greed. Lack of secure tenure may be a consequence of land ownership patterns, tradition, or civil unrest. Many developing country agriculturists have no adequate documentary proof of land rights, leaving them vulnerable to eviction, and consequently reluctant to invest in sustaining production. In many regions land reform is as important as agricultural innovation.

Well-managed soil can be a sustainable resource, and one that may even be improved; however, when misused the risk is that it will be degraded or even wholly lost. Soil degradation may be rapid and obvious – e.g. gullying, severe sheet erosion, soil crusts or hard layers – or it may be gradual and insidious. The latter is worrying because it can go largely unnoticed until a threshold is reached – the minimum depth to sustain crops or natural vegetation – and then there may be sudden disaster (Crosson, 1997).

A good deal of evidence suggests the world has had a serious soil degradation problem for decades (Eckholm, 1976). There should be plenty of hindsight knowledge, given the experiences of the USA Midwest Dust Bowl disaster of the early to mid 1930s and subsequent problems in many other parts of the world. Nevertheless, soil management has not had anywhere near adequate attention or funding. With perhaps 45 per cent or more of the world's land surface significantly affected by soil degradation, funding for soil studies, conservation and remedial activities are inadequate. Some governments have even cut back on their soil survey, monitoring and management over the last few decades; one reason is that some developing countries were forced to adopt structural adjustment programmes in the late 1980s, and these have led to cuts in environmental management funding and often trigger socio-economic changes which stress the environment.

Soils become degraded through many causes: they may be eroded – i.e. be scoured away by the wind or running water – they may lose their fertility through poor farming, overgrazing or pollution, they may become salinised (salt or alkali contaminated), compacted, develop a crust or impermeable layers, get waterlogged, their carbon content may be oxidised away, they can suffer acid deposition, or urban and other infrastructure development might remove or cover the soil. Even slight alteration of vegetation cover or drainage can trigger soil changes, possibly leading to permanent degradation. Natural soil production is usually a very slow process, taking at best many decades to produce a few centimetres, and demands a good cover of vegetation. Improvements might be speeded up if environmental managers could encourage the production of compost and the use of green manure.

Events off-site, such as regional pollution like acid deposition, global environmental change, economic policies, warfare, the arrival of refugees, and many other things, can affect soils. Soil management must be aware of such developments, which means there is a need for the sort of overview an environmental manager can offer.

Desertification is a term now widely accepted to mean a process that results in land degradation, which may be difficult to reverse. Desertification is seen as mainly human-induced (so it is dealt with here, rather than as an environmental threat in Chapter 8). Sometimes it may be mainly due to natural causes, and often these are triggers which catalyse human mismanagement. If the causes are physical then curative efforts are less likely to succeed.

Many developing countries have seasonal or periodic shortages of precipitation, often compounded by high rates of evapotranspiration, extreme temperatures and freely draining soils. Even in the humid tropics sandy soils can dry out fast if their cover of vegetation is removed. The consequence can be reduced infiltration, leading to erosion, flash floods and siltation where the flows slow.

Ultimately, conditions may become more or less permanently desert-like and difficult to rehabilitate: desertification is an ugly word for an unpleasant land degradation process and its associated human misery. Desertification, although widely used, is a term that is ambiguous, imprecise and emotive (Adams, 1992; Kadomura, 1997).

Worries about desert spreading in Africa were being voiced in the 1930s (Stebbing, 1938; Aubréville, 1949). Subsequently, there has been considerable debate over whether it is due to natural causes, human activity, or both combined. Nowadays, desertification is usually seen as the product of mismanagement of vulnerable environments. Common causes are: overgrazing, wood fuel collection, bushfires, salinisation, natural and human-induced climate changes and biological factors, like the introduction of new species. When faced by drought or desertification the first questions that should be asked are: 'Is it natural?' 'Is it exaggerated by humans?' 'Is it wholly due to humans?' If nature is to blame attempts to control the problem will probably be a waste of resources; however, if humans are part of or wholly a cause, control should be easier (Kassas, 1999).

Most now accept that desertification is the result of mismanagement of vulnerable land and is manifest as loss of plant cover and then soil damage (http://www.fao.org/desertification/intro.asp and http://www.ciesin.org/docs/002-193/002-193.html, accessed December 2003). The mismanagement may take many forms, ranging from overexploitation or introduction of organisms that damage the environment to pollution or fuelwood gathering. There is a tendency for land use to develop during a good rainfall period and fail to adjust when precipitation declines, causing land degradation, which may grow in severity to become desertification. Desertification is more likely in harsh environments like drylands and uplands where vegetation is under stress and soils once exposed are quickly damaged. While the claim is often made that deserts spread, most desertification occurs *in situ*. Sometimes the process may be natural, but more often environmental stress exposes human mismanagement. Desertification does happen in rich countries, including Europe and the USA; it also occurs in humid environments where soils drain freely – even Amazonia. In a number of countries regions with seasonal rainfall shortage have rising livestock and human populations, for a number of socio-economic reasons, including improved healthcare and the process of marginalisation. However, desertification is not always a Malthusian process, whereby population growth causes land degradation; it also happens where human populations are very low. It can result from ill-advised provision of wells, which prevent livestock from moving to avoid overgrazing, veterinary care may allow herders to accumulate too many livestock, or it may result from economic causes. Environmental managers must be cautious to ensure they correctly identify causes and do not waste resources treating symptoms or wrongly diagnosed problems.

Following severe drought and desertification in the African Sahel between 1969 and 1973, the UN called a UN Conference on Desertification (UNCOD) in 1977, which drew up a Plan of Action to Combat Desertification. Countries were then urged by UNCOD to make their own plans, but a quarter of a century on there has been poor progress. Various agencies and non-governmental organisations (NGOs) continue to make extravagant claims that as much as one-fifth of the world's land area is 'desertified'. In 1994 a UN Treaty on Desertification was signed by a number of countries, and the UN currently has a Secretariat supporting its Convention to Combat Desertification, which claims that over 1 billion people were affected (http://www.unccd.int/main.php, accessed February 2004). While there is a serious problem, the data to back up various assertions are rather questionable. Much of the anti-desertification effort has been 'whistle-blowing' and advocacy. Insufficient effort has been expended on reliable research, monitoring and the development of desertification countermeasures. There has been a failure of the signatories to UNCOD to spend anything like the sums they had pledged – between 1978 and 1991 only around 10 per cent was actually paid-up.

Drought is often a recurrent problem and one that can spiral into greater seriousness when there is poor land management. It is generally gradual enough in onset to give authorities a chance to respond, if they have been vigilant enough to watch for warning signs. Drought management is undertaken in developed and developing countries, and is a complex field, partly because politics are often involved: authorities tend not to want to admit degradation or failure to act, and sometimes ignore a 'backward' or 'wayward' region's plight. The need for drought forecasting and management is likely to increase as populations in vulnerable areas grow, as marginalisation takes place, and as climate changes. International overseeing of drought management is desirable to counter the aforementioned national shortcomings. Morton (2002) observed that a number of developing nations now undertake drought management on a scale that is generally overlooked, so much more is going on than was at first apparent. Some countries may try to disguise the seriousness of the problem and their attempts to cope with it. This is probably in part because it is an unfashionable 'top–down intervention' which could also be seen as admitting mismanagement and as evidence of unsustainability.

Response to drought can include de-stocking grazing areas, food aid, protection of incomes, alternative livelihood provision and emergency water supply; some of these must be applied with caution to avoid dependency or causing people to relocate. The provision of new water supplies needs particular care – there have been cases where this has caused herders to overgraze land nearby and initiate severe soil erosion. Quite a few anti-desertification or anti-soil degradation efforts since the 1970s have had little effect on the environment or local people's well being. It is not unusual in a region of overgrazing and impoverished agriculture to install a costly, 'high-tech' irrigation scheme producing export crops, doing little to improve local food supplies and providing little employment in what was promoted as anti-desertification development. And it may even have displaced locals into more marginal environments where they degrade the land. Some governments see marginal lands and their peoples as subversive or in need of aid to strengthen state hold over territory – fear of desertification is used as an excuse for increased controls from the centre. Desertification remedies have tended to be technical treatments of physical manifestations (symptoms); less often is there aid to counter human poverty or subsidies to counter unfavourable market prices, which might be actual causes.

Moderate estimates suggest around 80 million people are seriously affected by desertification, and affected areas have growing livestock and human populations so things are likely to get worse. However, the idea that there is a crisis of desertification must be treated with caution, for there are a number of ways in which environmental conditions and human welfare can be misread:

- As a consequence of misunderstanding physical, biological and human social conditions, through inadequate knowledge or cultural blindness;
- Through adopting a 'snapshot view': observations too constrained in time or space, or both; it is easy to overgeneralise about desertification or land degradation, and short-term trends may be interpreted as indicating longer-term patterns;
- By interpreting limited 'hotspots' as proof of a widespread problem;

- By using indicators that mislead, e.g. one study maps vegetation at the peak of a wet season or rainy climatic phase, another at the end of a dry season or during a dry phase;
- By mistaking symptoms for causes;
- In harsh environments where conditions fluctuate a lot of planners may mistakenly seek 'average' conditions, leading to mismanagement;
- An apparently resilient but vulnerable environment can suddenly be altered and that risk is often overlooked: this could happen when annual vegetation has set seed, which would germinate after the next rains, then one brief fire or spell of grazing at a crucial flowering time, or during regrowth, might break the cycle and initiate degradation;
- Periodic or random climate fluctuations may distort conditions – for example, an El Niño event might trigger a drought;
- A slight change may initiate a steady feedback effect: there are cases where vegetation clearance altered the albedo, made it hotter, and started a trend of drying that is difficult to reverse;
- By accepting received wisdom without checking its source.

Those dealing with land degradation, whether drylands or islands, must therefore be cautious; Lomborg (2001: 29) warned of relying on 'myths and rhetoric' rather than hard data. He noted that Easter Island is often cited, its history warning of a coming world environmental degradation crisis. The argument runs that Easter Island's isolated population exceeded the carrying capacity, could not escape, and fell into environmental and social degeneration – and this could be repeated on a global scale. However, Lomborg argued that of around 10,000 Pacific islands colonised by the Polynesians, only about a dozen suffered the Easter Island fate. A wide variety of people are involved with land degradation and desertification, and some have their own agendas: aid acquisition, local empowerment, regional vote-catching and so forth. One is well advised to heed Warren's (2002) warning that land degradation should be judged in a spatial, temporal and cultural context. Too much of what is published has not been based on reliable facts. Consequently, by the early 1990s a number of researchers were questioning the received wisdom on environmental degradation and desertification, notably: Thompson et al. (1986), Thomas and Middleton (1994), Leach and Mearns (1996), Fairhead and Leach (1996) and Lomborg (2001). Evidence has accumulated for some regions that challenges established understanding that serious degradation or even a crisis was underway (Ives, 1987; Ives and Messerli, 1998a, 1989b; Thompson et al., 1986). Reij et al. (1996: 3) observed that a crisis narrative suits some agencies and politicians, because it enables them to claim that their intervention is needed. There are also 'baseline concepts' used by ecologists, planners and managers that may be misleading. For example, carrying capacity is attractive as a benchmark – but exploitation which does not exceed carrying capacity may still damage a resource because some natural or socio-economic change introduces unexpected stress. Climax community and succession are concepts that have been attractive to resource managers, but are not fully researched and proven – hence when used to predict they may not give reliable results. Often environmental conditions are assumed to be more stable than they actually are. Common resources behaviour under exploitation is not adequately established. Improved vegetation cover is widely assumed to aid soil and water conservation, but some forms might not: a plantation of trees like teak (*Tectona* spp.) or eucalyptus may transpire more water than low scrub or grassland and can intercept fine precipitation and then release it as large and erosive drops from as much as 30 metres above the ground surface; the result can be decreased groundwater recharge and more soil erosion.

Ives and Messerlie (1989b) and others (Guthman, 1997; Niemeijer and Mazzucato, 2002) have questioned whether the widely recognised crisis of land degradation in the Himalayas was real; under careful examination of records some areas even appear to have been considerably revegetated since the 1930s, rather than having deteriorated as the 'myth' suggested. Fairhead and Leach (1996) doubted that the region they studied in Guinea was suffering degradation to the degree generally assumed by expatriates; in their view local people were sustaining or even improving the vegetation cover. For at least 40 years fears have been voiced that Sahelian Africa and other parts of that continent are in the grip of desertification, and that things are rapidly getting worse. This land degradation is also widely seen to be increasing the threat of global warming because it results in loss of soil carbon sequestration. The evidence to support this is still not conclusive. Recently, there have been reports that conditions in the Sahel have markedly improved, apparently due to more precipitation and because farmers have been adopting soil and water conservation – bodies like the UN Environment Programme (UNEP) should check their land degradation statements before going to press (Pearce, 2002a). Efforts have also often been wasted on insensitive schemes, which seek to impose terrace or check-dam construction to counter sheet erosion and gullying to improve crops; these may work in theory, but in practice farmers may not accept their value and fail to maintain them or extend their use. They may build a few so long as there is a grant or food aid to do so, and if not convinced of the value and that maintenance effort is worthwhile, they will abandon the attempt. Remedies must be sustainable, address causes, not just symptoms, and locals have to believe in the worth. So, if sheet or gully erosion results from rural depopulation, structural adjustment, market forces and so on, building check-dams is just treating symptoms. The real causes are better treated through economic policy changes, employment schemes or subsidies.

The keys to countering land degradation and desertification are broadly: improve vegetation cover and soil fertility, encourage soil and water conservation methods, support a shift to alternative land uses, which cause less damage, and try to establish alternative livelihoods to replace any that are damaging. The ideal is to seek long-term rather than short-term improvements, and to stimulate land users to initiate changes, wherever possible using their own labour and funds and building on established (traditional) methods to improve them. However, soil and water conservation efforts may not seem especially attractive to many poor farmers – the return may not be quickly apparent or obvious, and it may appear to benefit others rather than those paying for it. Development of new approaches to assessing the needs, attitudes and capabilities of local people, notably rapid rural appraisal and participatory rural appraisal, should assist efforts to counter soil degradation and desertification.

The food situation

Globally, food production has risen to an all-time high (World Bank, 2003: 84). In 1997, world food production, if divided on a per capita basis, could have given everyone around 2,700 calories per day – an adequate diet for most (Conway, 1997: 1). Yet in early 2001, food emergency situations arose in 33 countries and affected more than 60 million people. In Africa alone, 18 million needed food aid in 2001–2002, mainly due to drought and conflicts, but also through flooding in Malawi, Mozambique, Zambia and Zimbabwe. Recent estimates suggest that worldwide around 830 million people lack adequate access to food. About 210 million of these are in sub-Saharan Africa, 258 million in East Asia, 254 million in South Asia, with Latin America plus parts of the Caribbean and North Africa having about 8 per cent, 5 per cent of the total. Today people are hungry, not because there is insufficient food per capita, but because they cannot obtain it: either it is unavailable in their region or they are too poor to afford it. Women and children of the poorest social groups are especially prone to malnutrition and hunger (Pretty, 1999).

What can environmental management contribute to improving food supplies? A useful input is to offer a holistic overview sufficiently removed from actual production to be dispassionate. Archaeology strongly suggests that some pre-agricultural societies had more abundant and better balanced diets than many enjoy today, sometimes for much less labour input; and, because they could migrate easily, they were less vulnerable. Agriculture may be effective, when it functions well, feeding many more than hunter-gathering could, but it has drawbacks. Sedentary agricultural communities are also more likely to encounter some disease and pollution risks. There is little hope of establishing how often people starved before sedentary agriculture was adopted – it must have happened; however, once people farmed and were tied to one locality, famines and lesser degrees of hunger have been the lot of a significant proportion of people.

Hunger and famine are all too frequent in developing countries today, and by AD 2020 there are likely to be 2.5 billion more mouths to feed (worldwide). Hunger is an ongoing problem of too little food being available; famine is the catastrophic impact of hunger. Malnutrition is caused by a diet inadequate in some way: possibly insufficient quantity, or poor quality, or even an excess of some foods. The result of hunger and malnutrition are misery, possibly physical and mental underdevelopment, weakened health and sometimes death. Famine can result from socio-economic or environmental causes, or both. Often multiple factors trigger it and these may vary from event to event. Sometimes the problem is actual food availability decline – a shortage of food; alternatively, there may be supplies, but people cannot get access – because of poor transport; or they earn too little to pay the going price; or they are denied food for socio-political reasons.

The economist Amartya Sen studied how insufficient income can reduce people's entitlement (access) to food, even though supplies are present. He studied how in Bengal (India) in 1943 rural labourers received lower wages than urban workers did, so could not afford food at a time prices were rising, and starved. Getting food may be as much related to obtaining adequately paid employment or land reform as it is to improving harvests. There is also the challenge of providing better nutrition – especially improving supplies of protein for developing countries. When food supplies falter, food aid can consist of emergency rations, assistance with research, extension and training, help with infrastructure or development projects – ideally whatever is needed to restore adequate long-term production.

One definition of development could be 'seeking the situation in which a people no longer suffer severe hunger'. That has only happened in presently developed countries during the last 150 years or so, and here and there it has sometimes broken down during war or through other disasters. In the past some civilisations fed large populations adequately, in some cases for centuries. Environmental managers, briefed by environmental historians and other specialists, could warn present-day food supply bodies of challenges to stable food production and suggest ways of reducing risk. They could also encourage the consideration of alternative means of food production which powerful agribusiness is unlikely to initiate.

The reasons a growing number of countries have witnessed a decrease in hunger since the 1940s include: improved transport and storage; agricultural development, especially better seeds, chemical fertilisers, irrigation, and pesticides; and socio-political developments (Action against Hunger, 2001). However, a lot of the world's farmers are still poor; many live in harsh environments with bad communications and are in poor health through diseases like malaria. Their livelihood is subsistence agriculture or something similar, so they have little cash surplus to save against failed harvests or invest to sustain and improve production. Because such agriculturists have little money and grow low-value crops, agricultural industries tend not to invest in improving their lot because there is little profit in it. Consequently, commercial agriculture may be flourishing while close-by subsistence agriculture supporting considerable numbers of people is weak. Agroindustry, national development bodies and even international aid agencies may not see things this way, and it takes NGOs and environmental managers to stress the issue and suggest the alternatives. There has been a boom in soya production for export in Brazil – about 25 per cent of the world's crop is now produced there. This is in contrast to the limited progress in improving Brazilian peasant food crop production; the oilseeds may bring in foreign earnings, but there is no guarantee these will be invested to help poor people. Also, the spread of soya in Brazil, Bolivia and Peru, having converted huge swathes of savannah to monocropping for export, has forced poor families off the land, or into marginal environments. Soya now seems poised to decimate Amazonian forest biodiversity (Fearnside, 2001). Soya can now be commercially grown with pseudosymbiotic bacteria, which fix nitrogen and remove the need for chemical fertilisers; what might such improvements have achieved if applied to small farmers' food crops?

There is a growing trend toward monocultures (monocropping); these fields of genetically identical crops are generally more vulnerable than traditional genetically diverse crops, their mechanised production means less employment, and they can cause environmental and consumer health problems – a high price for cheap food. Modern agricultural developments are likely to give agroindustry a growing control over inputs and a strong control over production. This could drive many small farmers from the land and have adverse environmental impacts. Environmental managers at national and international levels need to be vigilant for these sorts of threats and lobby to try and prevent or mitigate them. Much of the soya exported from developing countries goes to developed countries to feed livestock, especially the European Union (EU), which has shifted from purchasing from the USA because its soya is likely to be genetically modified. In the EU livestock and cropping are increasingly diversified. Livestock manure is a serious problem and is generally not used as fertiliser; crops therefore get chemical fertilisers, and animal feed is imported. In the past most farms integrated cropping and livestock, which meant less impacts and no need to import feed.

It is often forgotten that affluent nations like the United Kingdom (UK) had famines in the past with much lower populations than there are today; for example: in AD 1321, 1314, 1302, 1294, and many more up to the late 1840s. The causes were one or more of the following: inclement weather, crop, livestock or human disease (which hit labour input), warfare, a need for land reform, unfavourable economic conditions and rising population. No country today should be complacent about food production and the stockpiling of adequate reserves.

Improving, developing, modernising and transforming are all terms applied since the 1940s to efforts aimed at upgrading agriculture. Until recently, yield increase was the main goal, and improved security, sustainability, equity and minimising unwanted environmental and social impacts had much less attention. For most investors maximising harvests is still the prime goal – increasing population also prompts efforts to increase yield. So environmental managers must encourage concern for sustainability, environment, security and other goals.

Agriculture is a complex and diverse process, or more accurately an ecosystem and socio-economic system, pursued via many different strategies. To succeed, each strategy has to effectively manage a complex mix of environmental, natural resources exploitation, cultural, political, social, technological, institutional, economic and legal issues. Those discussing agricultural development often start by attempting to draw up a typology or classification which helps deal with the complexity. Without going into detail about techniques, it is possible to divide them into sedentary and non-sedentary. Most of the world's agriculture is now sedentary, and relies upon the ongoing use of a given land area. Nevertheless, there are still large numbers of non-sedentary shifting agriculturists in some developing countries. With recent social, economic and demographic changes these people increasingly fail to sustain adequate production and must either find alternative livelihoods or make a transition to sedentary agriculture. Generally they get little help to do so.

The world's agriculture can also be crudely divided into family and non-family or subsistence and commercial. But a more relevant division might be into those willing and able to invest in sustaining production and innovation, and those unable or unwilling to do so. Where funds are scarce the latter might include commercial farmers as well as poor smallholders. Ideally aid should focus on those willing to improve agriculture, but who lack the funds to do so.

Before the 1930s, cereal yields in developed and what are now developing countries were roughly comparable. By the 1940s a number of richer countries had boosted their harvests through improved seeds, fertiliser applications and better techniques, including pesticide and herbicide use. This 'intensification' – heavier crops, multiple cropping, improved yields and possibly enhanced nutritional quality – was achieved through use of seeds, some with reduced growing seasons, allowing more than one crop a year, which responded favourably to inputs, the provision of irrigation, and agrochemicals. Since the mid 1960s the hope has been that, with the right aid, large parts of the world can intensify; this strategy has generally been called the Green Revolution. The alternative to intensification is 'expansion' – extending food and commodity production into new areas. In 1990 roughly 11 per cent of the Earth's land surface was cropped (used to grow food or commodities – this excludes grazing land), most lying within 35° of the Equator (Pierce, 1990: 35). Easily usable land is already occupied – further expansion will require breakthroughs in crop breeding and techniques that will overcome lack of moisture, saline soils, infertile land and pests. It may also require social changes and community developments.

There are situations where a population survives with a barely adequate diet, which renders it vulnerable to periodic severe hunger, and hinders labour input to improve things. Many societies have lifestyles that include seasonal hunger. Typically this falls at a time when the last harvest has been consumed and the next has yet to mature. Poor weather or other misfortunes can debilitate people so that they plant less and become ever more vulnerable – a vicious spiral of decline can then ensue (Chambers, Longhurst and Pacey, 1981). These people need better food security and perhaps occasional aid.

Food and commodity production means developing agro-ecosystems (modified natural ecosystems) to produce harvestable material. This demands inputs, various 'mixes' of labour, production skills, water, fertilisers, non-human energy, seeds, possibly livestock and feedstuffs, and disposal of waste salts, surplus agrochemicals, manure and food or commodity processing effluent. It is reasonable to say the world presently relies for most of its food on agro-ecosystems producing four species of cereal – wheat, rice, maize and barley. The most productive of those agro-ecosystems are limited in numbers and may not be sustainable under current practices. Humans, rich and poor, are too reliant on a limited range of food sources, with sizeable surpluses produced in too few areas. In 1971 around half the wheat imports to developing countries came from North America (pre-Green Revolution), in 1972 the Soviet Union was forced by poor harvests to buy up surplus grain stocks and in 1974 there were poor North American grain harvests; this quickly triggered a several-fold global food price increase. Since the 1970s, global human populations have grown, especially in developing countries, and cereal production increases have so far kept pace. However, there has been a trend toward more imports into developing countries of wheat produced in other developed countries. This is sometimes a cause and sometimes a consequence of falling cultivation of traditional crops. Water shortages and labour shortages have prompted food importation (see last chapter). There is also growing demand for wheat caused by westernisation and urbanisation – people are eating more of it. These food consumption and production trends are worrying; add to this the breakdown in modern agriculture which may occur because its inputs pollute the environment, and the possibility of global environmental change, and it is clear food security should be reviewed.

Food security can be defined in many ways, but a useful one is a mechanism that ensures people would not starve if faced by one or more years of scarcity. That means adequate surplus stored in more than one place and sufficiently dispersed to prevent total loss in a major disaster, and in sufficient quantity to cope with more than one failed harvest. History has shown that people can be hit by a number of successive poor or totally failed harvests. So security requires large quantities to be stored. Over 60 per cent of cereals produced by developing countries in 2002 was consumed by livestock, a large proportion of which are in developed countries. So finding surplus to store may be difficult.

Production and storage of food surpluses on a national or global scale is costly and presently is not encouraged by market forces or public opinion. Those who store food are mainly developed country or multinational commercial organisations likely to try to maximise profit in time of shortage. One alternative is to try to establish some local strategy: regional food reserve stocks, secure and sustainable livelihoods, local famine insurance schemes. Another possibility is to somehow encourage non-commercial storage. In recent years grain surpluses have been falling, storage has been cut back, there is more demand for grain for livestock feed in developed countries, and growing interest in agricultural feedstocks for energy or industry. These trends discourage production and storage of food-grain reserves. Between 1950 and 1976 per capita food output worldwide was increased by around 28 per cent, mainly by intensifying production (Grigg, 1985: 81). The challenges now are to reduce hunger in developing nations and ensure improvements are sustained with minimal environmental impacts, build in adequate security against natural or human-caused problems and stockpile supplies in case there are disasters. In 1996 the World Food Summit in Rome pledged to seek food security for all and a halving of world undernourishment by 2015. The task is not getting easier with growing populations, global environmental change, pollution, land degradation and difficulties obtaining satisfactory irrigation water (Cohen, 1996; World Bank, 2000).

Areas of East Asia, sub-Saharan Africa, Central America and the Caribbean face serious hunger now – about 30 per cent of the total world population was undernourished in 2001. Human immunodeficiency virus/acquired immune deficiency syndrome (HIV/AIDS) will soon infect over 36 million people world wide which will reduce labour availability and seriously disrupt family livelihoods, especially in Africa south of the Sahara, where some countries have infection rates of over 30 per cent (FAO, 2001a: vi). In 1999, world agricultural output increased by about 2.3 per cent with good harvests of cereal in the Sahelian countries, but growth slowed to around 1.0 per cent by 2001 with cereal demand outpacing production between 1999 and 2001 (FAO, 2001a: 3, 17). World cereal stocks (reserves) fell by about 7 per cent in 2000–1. World production of fish and shellfish increased between 1999 and 2001; however, some of those stocks were showing stress and it is doubtful such yields will be sustained.

Sub-Saharan Africa is where food production is lagging farthest behind population increase; between the 1960s and 1999 the continent's per capita food production fell by about 20 per cent and food imports increased (Goodman and Redclift, 1991: 155; Pretty, 1999). While grain prices may be rising, other food or commodity prices may be behaving differently – for example, coffee fell markedly in 2000 to about one-third of what it was in 1993. Where agriculturists produce such a commodity they must sell much more in successive years to obtain vital farming inputs and buy food, and may find it difficult. In such circumstances farmers and herders tend to neglect land management; they may migrate to find employment, leaving less able people to manage, or may undertake additional production of livestock, narcotics and so forth, which further damages the environment (Robbins, 2004). Similar impacts can be caused by withdrawal of subsidies, grants and low-cost loans – often prompted by structural adjustment programmes or by World Trade Organisation impacts.

Food and agricultural commodity prices are affected by a diversity of global, as well as local and national, factors – e.g. oil price rises are likely to boost cotton and natural rubber prices because competing synthetics are made more expensive. The Uruguay Round of the multilateral trade negotiations meeting at Marrakech in 1994 to discuss a General Agreement on Tariffs established the World Trade Organisation (WTO). WTO agreements have affected how countries can subsidise inputs and apply tariffs and other controls to support domestic production and discourage foreign competition; hopes for free trade and 'open' markets may have some unwanted impacts on agriculturists and their environment. Again, environmental managers can usefully oversee such developments.

El Niño and other natural disruptions

Food and commodity production is conducted in the face of constant environmental change: extreme weather events, climatic fluctuations, tsunamis, volcanic eruptions, the movement and evolution of pests and diseases, and many other setbacks. A significant portion of past famines can be blamed, at least in part, on such natural causes; the 1840s Great Irish (Potato) Famine possibly resulted from crop diseases prompted by climatic fluctuations (Davis, 2001). In the past few needed convincing that pests like locust and quela birds (*Quelea erthrops*, or the red-headed dioch) or livestock disease like rinderpest in Africa could spell disaster; however, effective pesticides, vaccination and other controls used between the 1940s and 1980s have caused a false sense of security. Recent warfare and austerity measures in many parts of the world have hindered monitoring and control of pests like locusts, and the cost of pesticides and stricter environmental pollution prevention measures have also made resurgence more likely. The outbreaks of foot-and-mouth and bovine spongiform encephalopathy (BSE) in the UK in the late twentieth century offer a warning of the need for ongoing vigilance, research, expenditure and effective quarantine (for statistics and maps showing pest threats see FAO, 2001a: 204–13).

Faced with a threat to food production or public health, and with a cost-effective countermeasure which could harm the wider environment or the population consuming produce, a developing country may take some convincing that it should not take risks. Well-informed environmental managers should review and steer such trade-offs. Palaeoecologists and historians have recently shown why the destructive North American locust (*Melanoplus spretus*) became extinct by the 1920s. The tillage of certain breeding grounds was the cause – and this might work for locusts elsewhere, saving pesticide use. Environmental managers can channel such information to appropriate agencies; and can recognise that it is also a warning that even limited development activities can suddenly have unexpected consequences: extinction of an abundant species.

The Green Revolution and beyond

The roots of the Green Revolution lie in the 1930s, when experiments and farm extension work to improve wheat and maize yields in Mexico were funded by bodies like the Kellogg, Rockefeller and Ford Foundations. The results by the early 1940s were good and stimulated the approach to agricultural modernisation described by the catch-phrase 'Green Revolution'. This was typically attempts to boost crop yields rapidly, without general social or economic reform, largely on existing land holdings, by making available new, potentially high-yielding seed varieties supported by cheap oil-based inputs of fertiliser, herbicides, pesticides and tractor fuel. The 'package' of new seeds, techniques and inputs could raise yields threefold or more, *if* everything ran smoothly. And during the Cold War, this was seen by many in the West as a way to avoid possibly violent land reform – a 'red revolution'. The Green Revolution, it was hoped, would feed the world and side-step demand for social reform.

It is possible to subdivide the Green Revolution into phases: the first occurred between the early 1960s and about 1974. The problems were many: the approach was insensitive to farmers' needs and abilities; the costs involved meant that poorer farmers could not adopt the 'packages' of innovations, while richer neighbours could, and the latter often bought out the former. Some of the improved crop varieties proved unsuitable in various ways. Pollution from agrochemical inputs caused difficulties and mechanisation meant some labourers were no longer needed. Successful Green Revolution adoption depended on correct use of a multicomponent package; any one of many things could go wrong and cause failure, so there were risks. Generally it was richer farmers could survive these and prosper (see Figure 4.1). Before the 1980s the improved seeds were high yielding varieties of rice, maize or wheat – so farmers reliant on root crops and other grains, like many in Africa, were by-passed for some time.

Around 50 per cent of developing country peoples were hungry in the 1960s – by 1974 this had fallen to about 20 per cent, and population had increased; so the first phase of the Green Revolution seemed to have staved off famine, and the balance of opinion is that it was an overall success. Critics, however, focus on the negative impacts and argue it failed and effectively ended in the 1970s.

A second Green Revolution phase can be said to have started around 1974, as a consequence of the Organisation of Petroleum-Exporting Countries (OPEC) raising oil prices, which affected key petroleum-based inputs, and because the negative socio-economic and environmental impacts were prompting efforts to find solutions. High-yielding varieties expanded from rice, maize and wheat to include some traditional cereals (like sorghum) and some non-cereal crops such as cassava

Figure 4.1 *Green Revolution. Intensive rice production based on high yielding varieties, chemical fertiliser, herbicides, pesticides, irrigation and mechanisation. Part of the Muda Scheme, Kedah, Malaysia.*

Source: Author, 1978.

(Lipton and Longhurst, 1989). The approach to implementation was becoming more sensitive and methods were increasingly appropriate.

Environmental impacts were serious enough by the mid 1990s to threaten further progress and some successful areas had to cut production. This prompted calls for greener approaches. So 1995 to the present might be said to be a third Green Revolution phase, aiming for a *Doubly Green Revolution*, which will raise food production sustainably and seek to avoid environmental damage.

The first phase Green Revolution was based on crops bred to respond to fertilisers and irrigation and produce more useful products. Unimproved crops would tend to respond to those inputs by growing more leaves which would weaken grain production, reduce its quantity and hinder ripening. Other yield-enhancing qualities included short growing season varieties able to provide multiple crops in a year; varieties with stems that resist damage in bad weather, or with multiple heads of grain and less leaf. By the 1990s new approaches included breeding disease-free stock – which could, in the case of potatoes, yams or cassava, give improved yields by reducing debilitation without other inputs or altered cultivation methods.

Today, it is clear that Green Revolution chemical fertilisers, herbicides and pesticides are causing environmental problems, and established land husbandry is poor at preventing soil degradation. All this makes sustainable production unlikely. Current irrigation and livestock-raising approaches are also emitting methane which significantly contributes to global warming; in 2010 this could account for as much as 40 per cent of total forcing of climate change. So, methane-reduction approaches need to be part of the Doubly Green Revolution.

The Global Land Assessment of Degradation (GLASOD) programme suggested in the 1990s that, since 1945, as much as 23 per cent of the world's productive land had been degraded to the point of uselessness. The accuracy of these estimates is still to be established, but that there has been much loss is beyond doubt. So future agricultural development has to avoid agrochemical pollution, ensure it does not raise greenhouse gas emissions, practise effective soil and water conservation, withstand global environmental change, be accessible to a wider range of farmers, offer more security – and increase yields further.

Much of the Green Revolution co-ordination and development so far has been through the Consultative Group on International Agricultural Research (CGIAR) and the International Agricultural Research Centres (IARC) network – a chain of specialist institutions working to improve crops, agricultural techniques, and ways to promulgate innovations and improvements. Ultimately the CGIAR is overseen by the FAO and OECD. For a list of Green Revolution institutions and details of their activity, see Conway (1997: xiv–xv).

Much of the Green Revolution literature is anything but analytical and impartial, and it is common for both supporters and opponents' moral judgements to be repeated as 'truths': small farmers are good, large agribusiness is bad. Environmental management must ensure developers see beyond such 'polarised perceptions' and rhetoric to back the approaches that give good, secure, environmentally sound results.

Doubly Green Revolution

Improvement of security of harvest, sustainable production and equity issues have had some attention. There is growing interest in approaches that seek sustainable rural livelihoods. To pursue the latter, multidisciplinary appraisal is used to identify the whole spectrum of needs, capabilities and hindrances which shape the rural livelihood strategies of a given locality and group of people; improvements are then sought which enhance human well-being, whilst reducing unwanted environmental and other impacts, and which can be sustained indefinitely (see Chapter 9 coverage of RRA and PRA). Clearly, this is a challenge, but it could offer means to secure food and dignity for people even in remote and harsh environments.

Where there is more money to invest, and supplies of inputs and marketing of produce permit, there is potential for large-scale and high-tech improvements to create sustainable food systems. Some environmentalists and development specialists oppose large-scale commercial production and technology, but to feed the world both large and small producers have to be supported and the benefits of technology will be needed.

For example, large-scale and small-scale intensive chicken rearing could use waste from pigs, and feed its waste to aquaculture – waste becomes a valuable input and pollution problems are at least part-resolved, although experiences with BSE in the UK and avian influenza in China urge caution. Integrated pest management and integrated plant nutrition approaches both have potential for reducing clumsy use of polluting chemicals. The problems involved in reducing environmental impacts, sustaining crops and raising yields are not the same when production is in the hands of marginalised agriculturists, with little if anything to invest, as they are for larger-scale producers (Power, 1999). At the larger scale, production is promoted mainly by commercial interests who seek to maximise short-term profits. Smaller producers are generally more concerned with security of production because they have little to fall back upon; also they need to see reasonable benefits from any labour they invest. Countries like China and Cuba have invested state support in large-scale and small-scale producers, and some innovations are not aimed at commercial advantage.

Jules Pretty (1999) reviewed the prospects for developing sustainable agriculture to feed Africa's growing population of small farmers and herders. He noted successes in Cuba which might be transferred, including organic farming approaches, i.e. reducing chemical inputs, and the improvement of social capital. Participatory approaches, rather than coercion, he felt promised to be the best route (Pretty and Ward, 2001). Agricultural improvements may not transfer smoothly from one locality to another, even if the environments and societies seem at first glance similar.

Silver Revolution

Worldwide there is growing pressure on marine and freshwater fish stocks. In some countries it is indigenous commercial fishing which is decimating stocks, but often the damage is done by foreign vessels, which are difficult to police. Some of the offenders are from developing countries and some from richer nations. Affluent countries with marine fisheries, like the Falklands, are able to invest in monitoring, policing and enforcement, but many developing countries simply cannot afford this (Moore and Jennings, 2000). In some cases poor nations have granted licences to richer nations' fleets to fish their waters (e.g. Morocco), and the stocks are being overexploited. Away from territorial limits ocean-going fleets are taking migratory species so that many countries now have declining catches. There is a widespread need for better monitoring and control of fish stocks if these are to be sustained. Uncontrolled fishing can also impact on 'non-target' organisms, notably turtles, marine mammals, seabed organisms and seabirds. Controls can have biodiversity conservation as well as stock management benefits. Where there are periodic environmental changes like El Niño events or 'red tide' toxic algae blooms, fisheries management must be especially vigilant to avoid irreparable decline in stocks. Unfortunately, international law and agreements governing fishing are far from satisfactory. Freshwater fisheries and inshore marine fisheries are often badly hit by chemical and sewage pollution, oilspills, silty river runoff, reduced nutrient flows because of dam and barrage construction, and the loss of breeding and nursery areas through river regulation and removal of mangroves and saltmarshes. There has also been breakdown of traditional taboos which once helped conserve marine and freshwater fish stocks. In the Southwest Pacific there have been moves to try and restore traditional fisheries taboos to conserve stocks and conserve biodiversity (Young, 2004).

Aquaculture offers the chance to augment marine and freshwater fish and shellfish production, which is essentially hunter-gathering, to intensive and potentially sustainable cultivation; this has been called the 'Silver Revolution'. Unfortunately, so far, most investment has been in low-tech and unsustainable production of luxury foods, notably tiger prawns; game fish, seaweed and oysters – for food and pearls. These are very profitable, but rely on unsound inputs and cause pollution, loss of common resource access for the poor, damage to biodiversity and other unwanted impacts. What is needed is a Silver Revolution that is doubly green – sustainable aquaculture that produces protein or other useful commodities for a wider market, with minimal environmental and socio-economic impact (Gavine et al., 1996).

Future agricultural development: Gene Revolution and Blue Revolution

The impacts of established Green Revolution approaches have already forced changes in land use after less than 30 years. The following problems are typical:

- More and more parts of the world have been forced by pollution caused by agrochemical use to set aside areas – i.e. convert agricultural land to a less-polluting use, such as from grain production to golf courses or forests.
- A number of sugar-producing areas have been linked to algae blooms in rivers, lakes and coastal waters; off northern Australia and Indonesia coral reef and fisheries damage has been linked to agricultural, and especially sugar production-related, agrochemical pollution.
- In India, states which have been at the forefront of the Green Revolution now have serious pollution which is forcing a reduction in agrochemical use – but the new varieties need those inputs.
- Groundwaters, rivers and wetlands in many parts of the world have excessive levels of pesticides and other contaminants.

Agricultural development must reduce costly and potentially polluting inputs and become more sustainable. This may be via enhanced nitrogen-fixation crops to cut the need for fertilisers, and developing crops which resist pests. Some of these reductions can be achieved through conventional crop breeding and biotechnology. Biotechnology is the utilisation of biological processes and the manipulation of living material *without* transgenic changes. Biotechnology includes fermentation, micropropagation or cloning plants from a fragment of living tissue, selective breeding and much more. Many of the approaches are long-established: plant and animal breeders have used selective breeding for thousands of years. There have been attempts to increase mutation rates to form new varieties since the 1950s, using gamma rays, x-rays or chemical treatments. However, progress relies on serendipity – a mutation may prove useful but many are not – so improvements are slow to achieve.

Faced with possibly rapid environmental changes and population growth it is vital to have ways of quickly producing new crops, livestock and pharmaceuticals 'tuned' to actual needs; this should be possible with genetic engineering which creates genetically modified organisms (GMOs), greatly expanding opportunities and speeding innovations. For example, pest deterrence can be introduced into crops by making their sap distasteful or toxic to pests. Or a crop can be created which is herbicide tolerant so weeds can be simply removed by spraying. Great caution must be exercised, for not only are there risks of contaminating biodiversity from GMOs (see Chapter 5), there is also a chance the innovations will misfire. In the USA, Argentina and some other developing countries there are already problems with glycophosphate-tolerant GMOs; these crops can be sprayed with herbicide, but weeds have already adapted, causing farmers to spray *more* weedkiller and contaminate the environment.

For poor farmers, especially in remote and harsh environments, the best way forward is probably through low-tech solutions, using local materials and human labour to intensify (see Box 4.1). Land use and water resources management must be better-integrated and sustainable agriculture improved (Calder, 1999).

Genetic engineering involves manipulation to produce recombinant DNA, which can be transgenic material incorporating DNA from widely different species (Shiva, 1993; Barrow, 1995: 212–15; the Institute of Development Studies has published papers on biotechnology and developing countries, available online at http://www.ids.ac.uk/biotech, accessed March 2004). Transgenic organisms have been produced since the 1980s. Huge debates currently rage over the production and release of GMOs. GMO (or GM) crops have already been adopted in many countries for food, fodder, pharmaceuticals and other uses. For example, bio-

---

Box 4.1

A promising strategy to raise poor farmers' agricultural production in African dryland environments

Cropping over considerable areas of seasonally dry Africa is blighted by weeds, notably *Striga* spp. These are parasites of cereal crops, which greatly depress yields. The traditional farming strategy is a form of bush fallow, i.e. non-sedentary shifting cultivation: a crop is grown, then the field is left out of production for some years to recover fertility and discourage weeds. So perhaps only one year in 10 or more yields crops. Attempts to reduce the fallow period are hindered, even if fertiliser or compost is available to maintain soil fertility, because *Striga* seeds are dormant in the soil or it infects certain weeds.

An improvement would be to seed between cereal crops with a suitable legume, possibly a crop or a green manure, which adds nitrogen and organic matter to the soil. This helps maintain fertility and discourages weeds that are host to the *Striga*. Some legumes also cause dormant *Striga* seeds to germinate and then die because there is no cereal to parasitise. Potentially it could allow a reduction in fallow time – and more crops. This may prove a sustainable strategy to boost crop yield without demanding much input of fertiliser or herbicides or a lot of additional labour. It should be easily grasped by small farmers and promises to be easily afforded.

Source: Several, including Pretty, 1999; Abunyewa and Padi, 2003.

pharmacy now produces much of the world's supplies of human insulin using GMO bacteria. The advantage is speed of production, large yields and flexibility and no animal cruelty (http://www.colostate.edu/progravis/lifescience/TransgenicCrops/hotbiopharm.html, accessed February 2004). Crops like soya and maize are increasingly grown with GMO seeds, which offer tolerance to herbicides or the ability to deter or resist pests.

The fear is that GMOs could go out of control and cause disasters: for example, a biopharmaceutical GMO could pass genes to food crops, which would then produce the pharmaceutics compound so people might find it difficult to avoid a drug or hormone intake. GMOs, unlike chemical pollutants, could continue to increase after escape; and because it is in breeding material a 'clean up' could be difficult. It could be a serious threat to biodiversity conservation because species might be corrupted by contaminant DNA.

In addition to fears of GMO accidents, there are problems of developing country access and dependency. So far, biotechnology, and especially genetic engineering, is largely in the hands of affluent countries and the private sector; although in China it is mainly state controlled and directed more to developing country needs. There is a risk that GMO advances could consolidate an already powerful handful of agribusiness and food companies. The implications are that it could make it difficult for poor farmers to compete, and research backing would tend to be focused on things likely to give profits to companies. Efforts need to be directed to applying GMO research toward pro-poor and environmental management goals (Avramovic, 2003).

GMOs could increase the speed, scope and precision of crop breeding and offer pollution-free alternatives to chemical fertilisers, pesticides and herbicides; it may be a way to enhanced photosynthesis, enhanced petroleum recovery, bio-treatment of waste, and much more. In Kenya the Insect-Resistant Maize for Africa (IRMA) project seeks to develop GMO corn resistant to stem borer insects. This has the potential to cut crop losses by 15 per cent or more and the seed should be accessible to small farmers. GMO cotton has been engineered to express insecticidal toxin derived from the bacterium *Bacillus thuringiensis*; it has been widely adopted by commercial growers and small farmers around the world, having met less consumer objections than food GMOs, and now accounts for over 20 per cent of the total crop (http://www.ids.ac.uk/biotech, accessed February 2004). If GMOs can enable cultivation of poor soils or the use of saline soil and salty water, huge expansion of cropping could be possible.

Environmental management must predict, monitor and reduce risks; it is also important to police GMO raw materials – germplasm from conservation areas, the general environment, and collections like botanical gardens or cryogenic gene banks. This 'policing' should ensure all legitimate researchers can access raw material for agricultural improvement, and demands checking that collections are openly and fully catalogued and that potential users can obtain information and samples. It may also be necessary to lobby and support collecting, research and development which could help the poor but yield little profit for biotech companies. Policing should also help ensure that commerce or terrorism do not misuse biotechnology, and that biodiversity collections are securely and adequately funded.

Collections should be sited where they are at least risk from environmental change, disasters, civil unrest or whatever. At present a number of biodiversity collections are vulnerable. Without such measures most improvements would be focused on commercial production and crops which benefit the more affluent.

Intellectual property rights (IPRs) are supported by biotechnology companies as a means to ensure they recoup the costs of research and development; however, their use may discourage non-commercial developments. The poor might gain from IPRs if they are used to generate funds, which can be targeted to their needs. Ideas for biotechnology development have often come from studies of traditional use of natural resources, but those offering the information seldom get any reward. Well focused, carefully monitored, needs-sensitive development of biotechnology could help solve the world's food supply problems, cut pollution and make production less vulnerable: a Gene Revolution.

To summarise: genetic engineering is a double-edged sword – it could offer great benefit, but there are risks associated with it:

1 Accidents where a modified organism or genetic material 'escapes' and damages some crucial component of the environment, or food production, or causes illness.

2 Poorer countries and fields which are not commercially attractive are likely to find it difficult to get funding for GMO research and development. Few developing countries have so far developed significant biotechnology capabilities and many could become dependent on corporations or developed countries.

3 Genetic engineering and biotechnology could quickly enable the substitution of truly natural products with industrial production of 'natural' products: chocolate, vanilla, vegetable oils, virtually anything. Some developing countries are very dependent on the export of such commodities and would suffer if big business undercut producers in this way.

There is deep public disquiet in many countries about the spread of GMOs. Keeping GMO crops out of food supplies for those who wish to avoid them can be difficult; already much of the world's soya is genetically modified and once stocks leave farms cross-contamination is a risk. At the time of writing Europe and the UK had a virtual moratorium on growing GMOs and their importation (http://www.gmsci.encedebate.org.uk, accessed February 2004). Reluctance to import GMO crops has meant a shift in purchasing, in the case of soya, away from the USA to countries like Brazil. An often-acrimonious debate is in progress between the USA, which is more supportive, and Europe, which is more opposed to GMOs as food crops. GMOs are offered as food and agricultural aid but some countries have voiced concern: in 2002, Malawi, Mozambique, Zambia and Zimbabwe refused or hesitated to take food grains. Some of those countries only accepted the aid on condition it was milled to prevent any chance of it being planted. There are also worries that GMOs will make growers too dependent on agribusiness. Already, a number of crops have been made resistant to certain weedkillers, which means the company has a captive market to sell herbicides as well as seeds, and it fails to capitalise on the GMO potential to reduce agrochemical use (http://www.ohiolline.osu.edu/gmo/a1.html, accessed February 2004).

Environmental management can help food and commodity production by establishing:

- What the current environmental situation is (structure and function, thresholds to monitor);
- What opportunities there are for expansion or intensification;
- How improvements can be made with minimal impact on environment and biota and most chance of sustainability;
- What natural threats exist and how to reduce vulnerability to them;
- How human impacts like pollution and global warming will affect agriculture and how these can be reduced;
- What is needed for institution building to support food and commodity production;
- What legal, extension and training, forecasting and monitoring arrangements are needed;
- What ongoing monitoring, early-warning and response measures are necessary.

Currently there is little strategic control over agricultural technology and GMO crops are widely grown in at least 16 countries. In February 2004 the UN introduced the Cartagena Protocol on Biosafety, part of the UN Convention on Biodiversity; this provides countries with an opportunity to assess the risks associated with GMO crops before allowing their import. However, this could be seen as trade restriction under World Trade Organisation rules – the USA has argued that is what EU restrictions have constituted since 1998.

How environmental management can support agricultural improvement and food provision

Environmental management can help establish maximum practical food production levels, model possible strategies for improvement and assess their impacts, identify bottlenecks and needs (Goodland et al., 1984; Ervin and Schmitz, 1996). Forecasting something as complex as agricultural production in not a precise art and probably never will be. But it should be possible to foresee and be better prepared for situations like that which led to the 1840s Irish Potato Famine or the recent UK foot-and-mouth and BSE livestock epidemics. Efforts are currently being made to predict likely effects of rising carbon dioxide levels and climatic changes.

Given that it will never be possible to foresee and avoid all disasters, environmental managers should lobby governments to duplicate and stockpile more food and food production recovery materials in secure sites. There should also be efforts to improve food production by small farmers in developing countries rather than allowing the current trend toward importing cheap grain from North America or other regions with regular surpluses.

Historical ecology, environmental history and palaeoecology have attracted more interest in recent years, and research that would have been treated with suspicion between the 1940s and 1990s as being 'environmental determinism' is now giving an indication of how environmental change might impact on the future, and a better idea of how human society and economics could respond to problems (Barrow, 2003).

Environmental managers can examine conflicts between agriculture and the environment, and other human activities, and suggest ways of reaching satisfactory trade-offs or solutions. Farmers often have conflicts of interest with conservation and environmental protection bodies. There is a risk that rural and urban people increasingly hold different views, and their access to media and government is also likely to differ. This may lead to neglect of rural needs and prioritising of city demands – and may become a problem in developed and developing countries. In developing countries there is often conflict between rural rich or commercial bodies and poor land users. Environmental managers should have oversight of stakeholders and how complex systems like agriculture function and be in a good position to advise decision-makers.

Commercial farmers often argue that their profit margins are so tight that there is little for them to spend on environmental care, and poor farmers patently have no funds. Already, environmental managers advise aid agencies on how to assist the poor to establish sustainable rural livelihoods, and frequently vet projects to reduce unwanted impacts. What is needed is the capacity to assemble appropriate tools into packages to suit specific situations, minimise environmental damage and sustain production, and then effectively promote these (Conway, 1997; Tait and Morris, 2000). To some extent the CGIAR did this during the later part of the Green Revolution. Environmental management can play a key role in reinforcing that. Reith (2001) suggested the use of industrial ecology (see Chapter 7) to explore linkages and develop sustainable agriculture strategies and ways of 'levering' one resource toward multiple gains, together with the use of ISO 14001 for evaluating the functionality and environmental impacts of strategies that are developed – at least for larger-scale farming.

Sustainable agriculture in developing countries

Environmentalists, planners, aid agencies and many others agree that it is important to promote sustainable agricultural production. But there is no consensus over the form it should take or the priorities (Pretty and Howes, 1993: 13; Barnett et al., 1995; Shepherd, 1998: 23–55). This is hardly surprising, given the wide range of interpretations of what sustainable development means and the diversity of existing agricultural conditions. Sustainable agriculture has been defined as that which is 'ecologically sound, economically viable, socially just, humane and adaptable' (Reijntjes et al., 1992). I would add: 'integrating biodiversity conservation and soil and water conservation'.

Sustainable agriculture will take a diversity of forms: some will be mainly subsistence, some export oriented without being large scale and commercial – making use of strategies like fair trade to compete with bigger producers; and there will be large commercial producers. Most of the efforts to develop agriculture since the 1930s have aimed at boosting yields and sustainability, and social justice issues have had little support; for example, a widely used text on agricultural modernisation in the early 1980s contains no mention whatsoever of sustainable development (Arnon, 1981).

Sustainable agriculture has had some support from bodies promoting chemical-free organic farming and permaculture. In the USA there have been experiments comparing organic farms with non-organic; and in the UK Rothampstead Experimental Station has scientifically monitored areas of agricultural land for over a hundred years. There is also soil and water conservation promotion and research by bodies like the Soil and Water Conservation Society (USA) and various departments of the FAO.

Sustainable agriculture in developing countries must cope with a wide range of environmental conditions, frequently tropical with irregular precipitation and high temperatures. There is often no cold season to aid control of pests, soils are sometimes deeply weathered, nutrient poor and may have high aluminium or iron content. Added to these hindrances, many farmers can afford little for inputs and have poor communications with the rest of the world (Edwards et al., 1990). In a good number of regions existing agriculture is breaking down, as development pressures disrupt traditional strategies, or because of pollution by acid deposition, accumulating agrochemicals, or salts. So modernisation may have to take place from a foundation of degenerating practices and possibly failed attempts to improve agriculture (Pretty and Howes, 1993; IIED run a Sustainable Agriculture Programme focused on developing countries, available online at http://www.iied.org, accessed January 2004).

Sometimes sustainable development will come from outside, sometimes from within using ideas from indigenous peoples, women and small farmers, and in all probability there will often be a mix (Ghai and Vivian, 1992; IUCN, 1997). Change must be attractive to win wide adoption, so it is important to offer visible economic or labour-saving benefits as well as being environmentally sound.

Bodies like the FAO are promoting sustainable development, but there has been criticism that efforts are inadequate and sometimes a false facade (Shepherd, 1998: 41). There is little likelihood that agricultural improvement and a shift to better environmental care will take place fast enough, if at all, without environmental managers prompting and steering. Green aid might be a catalyst in some cases (a rather dated set of case studies is offered by Conroy and Litvinoff, 1988). In some situations it may be possible to diversify agriculture to include activities like ecotourism in order to provide a satisfactory livelihood and take pressure off the land. Much of the gain in production of the last 40 years has been through advances in crop breeding, fertiliser use and irrigation, at the expense of the environment. It may be that growing fertiliser use is masking increasing soil degradation – a sort of time-bomb effect (Fricker, 2000). What is needed is careful and appropriate fertiliser use.

The way forward for commercial growers could be via subsidies and tax incentives for sustainable development. Small farmers have little surplus income to invest and may be reluctant to take land out of production in order to improve soil and water conservation. They are unlikely to respond favourably to coercion and must be helped to see the need for sustainable activities and supported to pursue them.

One promising way to help sustain production is to use reduced tillage techniques. Zero tillage involves little or no ploughing and minimal soil disturbance, and can save moisture and reduce soil degradation. Controlling weeds and planting seeds present some challenges but effective methods are available. Between zero tillage and normal tillage are a wide variety of conservation tillage methods which slow runoff, trap organic debris and retain moisture (Barrow, 1999b: 25–7). Some soils are unsuitable for minimal tillage and are better managed with deep ploughing but many regions could benefit. There is a gulf between having promising strategies in research stations and selected farm trials and getting widespread adoption – this is true for all sorts of sustainable development strategies and tools, not just tillage. Strategies and crop mixes need to be assembled for each situation, tested to see that they work, are accessible and attract farmers, and then they have to be promoted, perhaps with limited outside aid (Barrow, 1999a, 1999b; Reij et al., 1996; Whiteside, 1998).

These challenges are often termed *transformation* – seeding with successful methods which become self-supporting, and hopefully spread of their own accord (Francis, 1994). As discussed earlier, one hindrance is the ready availability of cheap imported grain, which tends to drive down crop prices. Hindrances to better soil management are listed in Box 4.2. Environmental managers should promote site-adapted environmentally friendly 'ecofarming' and encourage different specialisms like soil science, agriculture, moisture conservation and so on to work in a more integrated way.

Box 4.2

What inhibits sound soil management?

- Soil can be rapidly damaged and is slow to form;
- Lack of soil specialists;
- Lack of funds;
- Insufficiently integrated approach to soil degradation;
- Soil degradation can be 'covert' – insidious and difficult to recognise;
- Degradation can be sudden, episodic and unpredictable;
- Since the 1940s development thrust has focused on crop yield increase and soil degradation has been neglected;
- Some impacts of soil degradation are delayed – current land users and investors seek profits in the near future, land degradation is a problem for people in the future and elsewhere;
- Soil conservation has a vague status – some academic institutions and government bodies undervalue it;
- Soil is affected by many things – tracing the causes can be difficult;
- Soil management is complex and has to integrate with livelihoods, political economy, and so on.

Source: Author.

Pollution caused by agriculture

Agriculture causes problems through pesticide and herbicide application, eroded soil, wind-blown dust, methane emissions from irrigated fields, fertiliser use and growing herds of ruminants, and in some regions smoke from fires set to clear brush or forest. Some pollutants travel great distances: dust generated by tillage in Asia is monitored in the mid-Pacific, and Saharan dust turns alpine snow pink from time to time and can reach the Caribbean. In the 1930s dust from the Midwest Dust Bowl soil erosion disaster blew over Washington and probably helped convince the US government of the urgent need for rehabilitation.

Pesticides

Pesticides are a problem, at local, regional and global levels. Some persistent compounds are now global contaminants. However, they have become a key element in disease control and maintaining food production, and until safer, more effective and affordable alternatives are developed, use will continue. There are quite high and little publicised numbers of deaths and illness due to pesticides amongst agriculture and food industry workers in developing countries. Consumers are also affected by contamination of produce so it is in the interests of developed country consumers to seek controls in developing countries.

Currently there is awareness of the problem and a lot of research and development work is focusing on integrated pest management: various approaches that avoid poorly-focused pesticide use, biological controls, pheromone traps and pesticides that affect only the target organism. In the meantime, developing countries and NGOs need to monitor commercial pressures on farmers to adopt pesticides that may not be appropriate and ensure that users apply adequate management practices. In particular there should be vigilance for misuse such as the mixing of 'cocktails' of a number of pesticides, over-frequent spraying and spraying in a manner that contaminates streams, crops and wildlife more than necessary. Considerable stocks of old and unsafe pesticides are in circulation and efforts should be directed to finding these, compensating the owners and safely destroying the material.

Malaria control and locust control still use pesticide. Shortage of funding, concern about using pesticides like DDT and unrest has hindered control of locusts in Mauritania and the western Sahara. So, while pollution has declined, there is a growing risk of locust resurgence in Africa. Alternatives like ploughing breeding areas demands political stability and funding – pesticide spraying by aircraft is often cheaper, quicker and avoids banditry.

Chemical fertilisers

Fertiliser contamination is a problem in many regions, and streams or groundwater carry the problem across national boundaries. The need to maintain, and if possible boost, food and commodity production makes it unlikely that there will be a rapid decrease in chemical fertiliser use; trends suggest the opposite. Yet in some places contamination is already threatening cropping. What are needed are suitable slow-release fertilisers, or other means which prevent rapid leaching and runoff, so that fewer and lighter applications are enough; this also cuts farmers' costs. There are other ways of cutting chemical fertiliser application, such as enhanced biological nitrogen fixation, green manure and compost application.

Both nitrate and phosphate fertilisers cause pollution of groundwaters, surface water bodies, streams and the marine environment. Most farmers apply a mix of these. The production of these agrochemicals also impacts on the environment: a number of developing countries supply phosphate, which can mean they suffer mining impacts and dust pollution; nitrate fertiliser production generally demands hydroelectric generation, which means the problems of large dams, and chemical processing of petrochemicals which wastes finite resources and causes pollution. Countries without fertiliser factories become dependent on overseas suppliers, which may not provide material when it is needed for endangered crops, and have to expend foreign exchange.

Where there are cities use may be made of sewage and refuse, if it can be processed to make safe compost. For many farmers the most promising option is green manure: crops, which can be ploughed in to enhance soil nutrients.

Low-cost pollution control and reduction of chemical fertiliser usage

Streams polluted with excess fertiliser may be channelled into aquatic plant-filled lagoons – suitable plants include water hyacinth, reeds or various algae. These absorb the pollutants and can be regularly harvested for compost to help sustain agriculture and reduce the need for chemical fertilisers, or as fuel, manufacturing or building material. The practice could cut downstream chemical oxygen demand (COD), biochemical oxygen demand (BOD) and sediment pollution. Silt trapped in lagoons can be periodically dug out and used as compost. Such lagoons are cheap, demanding little other than local labour.

Improved low-cost soil and water management has the potential to improve moisture retention, cut erosion and reduce the need to apply chemical fertilisers. These measures would help reduce the silt and nutrient pollution of streams. Even simple stone-lines along contours or crude terraces and rows of grass can effectively trap twigs, leaf debris, animal droppings and organic matter that would otherwise wash away or be scattered and oxidised – and so form regularly renewed and fertile soil where it is needed. Some approaches like stone-lines need little investment of labour and do not take much land out of production. Other forms of terracing need considerable labour to install and maintain and may reduce the cropped area. Sometimes authorities have installed these using tractors, but failed to convince local users of their value. The result is neglect, breakdown and locally severe erosion where terraces fail.

Efforts should be made to promote the use of organic composts and green manure – plants grown and then dug in to provide soil nutrients – the use of treated sewage and refuse (garbage) from towns (if free of heavy metals, disease organisms and other harmful contaminants). Another fertiliser reduction strategy is to use biological engineering to give crops the capacity to fix nitrogen, or to seed planted areas with suitable soil bacteria, which assist plants to do this. In many developing countries livestock manure is scattered and wasted; stall-feeding or tethering and feeding would allow the dung to be concentrated on cropped areas.

Some of these approaches are already established: in many rice-growing areas, farmers encourage the growth of nitrogen-fixing weeds and algae in the flooded paddy-fields, especially the aquatic fern azolla (*Azolla* spp.). New World farmers long ago developed crop mixes that enhanced nitrogen fixation: notably maize, beans and squash – the beans being the soil-improver nitrogen-fixing crop. Many areas could make more use of fertile silt-rich floodwaters, by spreading these to sustain farmland and cutting the need for fertilisers. Often all that is needed for this are cheap, easily constructed earthen-bunds (low banks).

Recent studies in Amazonia, following on from research by archaeologists, have shown widespread black-earth (*terra preta*) soils, evidence that in the past large communities were sustained on poor soils in harsh environments. Ancient peoples appear to have had a method which developed soil and retained nutrients even in 'infertile' areas – possibly based on the addition of charcoal and certain bacteria to the land. Studies in Brazil aim to discover how the strategy worked – success would offer the world a low-cost route to a Doubly Green Revolution. Archaeologists have also shown that large cities were once supported in Central and South America with raised-bed cultivation in a number of less-than-favourable environments. The methods might be promoted today as another doubly green and accessible strategy (see *chinampas*, Chapter 7).

Soil and water conservation and non-chemical fertiliser strategies deserve much support. Chemical fertiliser manufacturers may offer some resistance to this; also, some developers are familiar with fertiliser use – it offers them quick yield increases, whereas the damage it may cause through pollution and erosion is less apparent.

Concluding points

- Food production appears to be relatively insecure; much development since the 1940s is probably non-sustainable. Agriculture faces serious challenges and demand for food is also going to increase.
- Environmental management should advise agriculture on how to improve sustainability, and on whether strategies are vulnerable.
- The challenge is to promote approaches which raise yields, sustain production, reduce vulnerability and avoid environmental damage, and which are widely accessible.
- Land degradation is a threat to agriculture in many regions. Environmental managers should seek to ensure that it is correctly understood, and that remedial measures do not focus on symptoms rather than causes because of poor data collection and misinterpretation.
- The development of GMOs could offer ways of exploiting opportunities and countering problems far more effectively and swiftly than ever before. On the other hand, GMOs might run out of control and cause serious problems. GMO usage must be subject to the precautionary principle and very rigorously monitored by authorities

with a sufficiently broad view to be able to anticipate difficulties. Environmental managers can play an important part in controlling and focusing the use of GMOs.

Further reading

Conway, G. (1997) *The Doubly Green Revolution: food for all in the 21st century.* Penguin, London. [Argues that improvement of food production must show concern for environment.]

Madeley, J. (2002) *Food for All: the need for a new agriculture.* Zed, London. [Lively and readable coverage of food supply and agriculture.]

Thomas, D.S.G and Middleton, N.J. (1994) *Desertification: exploding the myth.* Wiley, Chichester. [Excellent examination of the desertification processes which questions dubious assumptions and excessive claims.]

Davis, M. (2001) *Late Victorian Holocausts: El Niño famines and the making of the third world.* Verso, London. ['Neo-environmental determinism' – explores the role of environment, and particularly El Niño/Southern Oscillation (ENSO) climatic events, in human history, especially the issue of famine. Argues that colonialism made some countries more vulnerable to natural disasters.]

Goodland, R.J.A., Watson, C. and Ledec, G. (1984) *Environmental Management in Tropical Agriculture.* Bowker, Epping. [The application of environmental management to agriculture in tropical environments.]

De La Perriere, R.A.B. and Seuret, F. (2000) *Brave New Seeds: the threat of GM crops to farmers.* Zed, London. [Reviews the problems associated with GMO crops and calls for worldwide safety measures.]

Leach, M and Mearns, R. (eds) (1996) *The Lie of the Land: challenging received wisdom on the African environment.* James Currey, London/Heinemann, Portsmouth (NH). [Questions the widespread incautious use of received wisdom on land degradation.]

Websites

Food and Agriculture Organisation (FAO): http://www.fao.org

International Center for Agricultural Research in the Dry Areas (ICARDA): http://www.icarda.org

International Crops Research Institute for the Semi-Arid Tropics (ICRISAT) – agriculture improvement in the dry tropics: http://www.icrisat.org/web/index.asp

International Water Management Institute (IWMI): http://www.cgiar.org/iwmi

International Soils Reference and Information Centre (ISRIC) – world soils and sustainable land use information: http://www.isric.org/index.cfm

Consultative Group on International Agriculture (CGIAR) – supports 15 centres around the world which have promoted the Green Revolution and continue to work to improve agriculture: http://www.cgiar.org/

(All accessed December 2003)

5 Biodiversity resources

Key chapter points

- This chapter explores the value of biodiversity and the threats to it. The various ways in which conservation can be undertaken are explored, and the consequences of not effectively doing so are discussed.
- Biodiversity conservation is underfunded, often poorly planned and badly managed. Conservation areas and gene banks are commonly unsustainable and vulnerable.
- Environmental managers should promote the funding of conservation, and ways of reducing the vulnerability of biodiversity collections. Conservation must be viable in the long term and collections need to be accessible to all legitimate users.
- The value of participation and involvement of local people – participatory conservation – is examined.

Biodiversity value and loss

Biodiversity is the diversity of different species together with genetic variation within each species in a given area. Myers (1985) called it the 'primary source', and it is material vital for sustainable development. It is vital to improve biodiversity conservation because:

- Crops, livestock and pharmaceutical products are constantly challenged by a changing environment and evolving pests and diseases, and must satisfy new demands and fashions; biodiversity is needed to meet these challenges.
- Without the 'raw' genetic material from plants and organisms of all sorts there will be great difficulties developing new crops, pharmaceutical products (drugs and antibiotics), new fish suitable for aquaculture, improved and novel livestock, bacteria and yeasts for fermentation or biotreatment of waste or composting, and many other innovations.

- Biodiversity has philosophical and aesthetic value and may inspire new ideas and scientific advances (Posey, 1999).
- A case can be made that humans have a moral/ethical obligation to protect biodiversity.
- Sustainable development dictates that humans should pass on to successive generations at least the same amount of biodiversity riches they have enjoyed.
- Biodiversity-poor environments may be less resilient and less able to recover if disturbed. However, there is considerable debate on this issue (Tisdell, 1999: 38–40).

Human knowledge is full of gaps; we are not sure what may be needed to ensure a stable, secure and sustainable future environment and human well-being. Biodiversity cannot be recreated once lost; it is a rich resource which should be held in trust for the future. For some time biodiversity losses have exceeded natural rates, and it is getting worse – to the extent that many argue that the world's greatest mass-extinction is under way and that humans are responsible.

Modern routes to improved agricultural productivity have mainly been via reduced biodiversity: the replacement of natural vegetation with areas dominated by very few crop species. Human activities are also reducing the diversity of insects, weeds and other organisms. Large agro-industrial corporations promote fewer crop varieties, and these are often bred in such a way or genetically engineered so that there is no chance of farmers saving viable seed. Growers become more dependent on business, and it is causing a reduction in food security.

For most of the time that humans have practised agriculture, domesticated varieties have grown in fields with wild ancestor species nearby, so that there has been ongoing cross-breeding which generates useful new varieties and maintains genetic diversity in the crops. Today farmers are compelled by commercial forces and legal controls to plant a limited number of varieties, so older types are lost while environmental degradation depletes the surrounding pool of wild species. Agricultural seed and biotechnology companies are effectively asset-stripping biodiversity from the wild and, in some cases, from the seed banks. There has been debate about the morality of this conversion of public resources to private. Morals aside, it is clear that companies that develop new crops and biotechnology seek to recoup their investment in research by patents and similar measures aimed at protecting intellectual property rights (IPRs) and by restricting access to their gene banks (Khor, 2004). These companies are mainly large corporations and their source of genetic material is often developing countries.

Many of the varieties of food crops in the USA have already been contaminated by DNA from genetically modified organisms (GMOs). There is a risk that this could take place worldwide. It could also affect material in gene banks because viability declines and stored material is regrown and seed re-harvested at intervals; during regrowth seeds could be contaminated by pollen on the wind or carried by insects (Pearce, 2004). In 2003 the Cartagena Protocol on Biosafety came into force; a legally binding international agreement which governs the transboundary movement of living GMOs (available online at http://www.biodiv.org/default.aspx, accessed March 2004). Hopefully, this will help counter GMO threats to biodiversity.

What is needed is a reliable open-access system of biodiversity conservation, an improved version of that which has been maintained by the international institutes funded by the World Bank, charities and some governments, which led the Green Revolution. Unfortunately, the latter are suffering shortage of funding, and even important botanic gardens like Kew struggle for adequate resources. What is needed is international agreement to create a permanent, legally binding biodiversity conservation system, which involves all countries. Failure to do so could result either in increased plunder of developing countries by powerful corporations seeking excessive control of genetic resources, or each country seeking strict sovereign rights over its biodiversity, which it can then sell to the highest bidder. Recently there has been lobbying to provide some rights for the source areas and peoples who have traditionally used the genetic material (Posey, 1990).

The Convention on Biological Diversity was negotiated just before the 1992 Rio Earth Summit, and came into force in 1993. It accepts that those countries with rich biodiversity would conserve their resources better if they could make money from it. As well as encouraging conservation the Convention supports sustainable use of biodiversity and equitable sharing of benefits (McConnel, 1996; Convention on Biological Diversity, availiable online at http://www.nhm.ac.uk/science/biodiversity/cbd.html and http://www.biodiv.org/doc/publications/guide.asp, accessed January 2004). In some cases trade in valuable species might be used to pay for conservation; the terms of the Convention on Trade in Endangered Species need reforming to achieve such ends, and to improve its effectiveness (Hutton and Dickson, 2000). In spite of sluggish support from the USA, Canada, Australia and New Zealand, there have been efforts to agree that in return for open access to biodiversity, centres would ensure their collections were freely available.

Often the progenitors of domesticated species of plants, livestock, fish and so on have restricted distributions in the wild, and can easily become extinct as land is developed or rivers and water bodies polluted or drained. For example, a large proportion of the world's total cocoa diversity is found in a relatively limited area of the upper Amazon where it is vulnerable to oil exploitation, logging, squatter settlement and forest fires. Roughly 40 per cent of the world's remaining tropical rainforest lies in the Amazon Basin; in the last few years Brazil has invested large sums in highway paving and new infrastructure projects, notably the Avança Brasil programme. These and the expansion of crops like soya are likely to accelerate biodiversity loss (Laurance et al., 2001).

How can remaining biodiversity be best protected? The main ways are to promote conservation (see Box 5.1), and to discourage property rights, trade practices, fashions and investment which act to destroy diversity (Frankel et al., 1995; Dobson, 1996). In addition, efforts should be made to promote awareness of environmental changes which might affect conservation areas and to counter any pollution which threatens biodiversity.

People are subjective in selecting what to protect. An organism may be saved because it is perceived to have potential economic value, or it seems deserving (Shiva et al., 1991; Perrings et al., 1995), or conservation occurs because a site or the

Box 5.1

Biodiversity conservation options

In situ

- Currently unexploited reserves and parks which have been little disturbed at any point, or areas once disturbed and now recovered from human influence. The latter are unlikely to have as great a biodiversity as the former, but nevertheless can be valuable.
- Currently unexploited reserves and parks which have at some point been disturbed and are some way off from recovering, such as regrowth forests and areas exploited in the past and abandoned to nature – old quarries, mined areas, abandoned farmland, military training grounds or minefields.
- Currently lightly exploited reserves and parks that are mainly used for conservation, but with limited human activity and extraction of products.
- Currently more heavily exploited areas which try to support some conservation, such as tolerant forest management, extractive reserves, recreational areas like golf courses, tourism areas, farming areas, agroforestry and road or rail verges.
- Artificially planted areas which may conserve some biodiversity – tree plantations which might be modified to offer more tolerant wildlife some refuge.

All of the above may be vulnerable to pollution such as acid deposition or stratospheric ozone loss-related UV, global climate or sea-level change, warfare, bushfire, etc. Ideally, these areas should be as large as possible, straddling enough range of environment to allow adaptation to changes in the long term. Buffer zones may be needed to protect crucial areas and care must be taken to monitor biodiversity to ensure there is no long-term decline – probably not something to be left to the conservation agency, which may be reluctant to admit failure. Some conservation areas can be credited against global warming pollution quotas, which offers additional benefits to conservation.

Some migratory and nomadic species, such as migratory birds, bats and insects, ocean fish and marine mammals, migrate or disperse freely so it is difficult to keep them in a single conservation area. Migratory birds may need a 'corridor' of reserve sites and protective legislation and refuge/ resting areas along flight routes. The conservation of migratory and nomadic fish, bird, reptile and mammal species, such as tuna species, marine turtle species and marine mammals, may depend more on internationally agreed control of hunting than conservation areas.

Ex situ

- Gene banks in the form of germplasm: dried seed collections*, chilled seed collections or plant tissue in deepfreeze or growth medium, deep-frozen animal sperm;
- Growing collections at dedicated sites: botanic gardens, zoos and arboreta*. These are not likely to have the degree of diversity that source areas have.
- The above are limited in number and mainly controlled by larger international, state or NGO bodies; ideally they should be duplicated in as many secure sites as possible and publish accessible lists of holdings. There is also a need for very careful quarantine of additions and thorough monitoring for any disease or pests which might destroy a collection.
- Smaller sites, individual farm or household garden conservation, city parks, local plantings. There is much scope for NGOs to assist with co-ordination and support: e.g. some seed companies encourage gardeners to buy and grow seeds of wild species which have been harvested without endangering natural stocks, individuals may keep various animal species, encourage wildlife in gardens and so on.
  * These are likely to need regular germination, regrowth and new seed collection at appropriate intervals before viability falls off. Regrowing means these plants may be periodically exposed to contamination from related species and GMO plants.

Source: Author

biota are sacred (Berkes, 1999). As Tisdell (1999: 27) noted, the loss of the dodo is often mourned, but the loss of numerous other species less well known or spectacular has had minimal public attention. Those seeking funds for conservation may find it less easy for species with less clear economic value or lacking attractiveness to the public. Where biodiversity is clearly seen to have the potential to generate future profits, commercial interests may undertake conservation. For other species, various forms of aid will play a crucial role in conservation.

Reasonable stocktaking of biodiversity resources is an important first step (Groombridge and Jenkins, 2002). If possible, an assessment should then be made to determine what is vulnerable (Heyward, 1995). Next, efforts should be made to identify the root causes of loss; these are often indirect, cumulative and complex, and so difficult to trace (Wood et al., 2000). Once all that has been done it should be possible to develop a biodiversity management strategy (O'Riordan and Stoll-Kleemann, 2001).

The fight to conserve biodiversity has intensified since the mid 1980s, and there are now many specialist NGOs which focus on specific ecosystems or species, as well as those with more general biodiversity conservation interests. A body active in seeking to conserve tropical rainforests is the World Rainforest Movement, an international coalition of NGOs that has been active since 1987. Biodiversity conservation was given centre stage when the World Bank launched a Biodiversity Action at the Rio Earth Summit in 1992. But much more still remains to be done.

Often conservation and livelihoods appear to conflict: efforts to protect forests may be greeted by locals involved in the timber industry as a threat to their livelihoods; attempts to restrict fishing methods or access to areas that have been traditionally used can result in opposition. Where conservation areas are seen to harbour pests or diseases that affect surrounding areas people may object and even trap animals. Large animals like elephants, large cats and those that migrate cause special problems because of their mobility and because some are a threat when they make forays from reserves into surrounding areas. Marine mammals and migratory fish also pose conservation problems because they need global protection.

Many conservation authorities argue that there is a need to involve local people and gain their support, to ensure they are protected from nuisance or dangerous species and, wherever possible, to offer them livelihood opportunities associated with conservation (see Figure 5.1) (Munasinghe and McNeely, 1994; Borrini-Feyerabend, 1996: Lewis, 1996; Jeffery and Vira, 2001; Koziel and Saunders, 2001). If not involved locals may feel alienated and then poach or resist conservation in other ways. Local people are increasingly involved in policing, administering and servicing conservation areas; they can bring to bear considerable traditional knowledge and are adapted to local conditions. There are, however, some very strong critics of the present fashion to integrate conservation with economic development and the encouragement of local participation. In West Africa such approaches may not be as successful as proponents claim, and Oates (1995, 1999) argued that conservation should be for the intrinsic value of the biodiversity, and not part of development. Clearly, participatory conservation initiatives need to be carefully planned and monitored, and not simply established according to current development or aid agency fashions and then left unsupervised.

Figure 5.1 *Green tourism. Tourists help pay for conservation and provide local employment opportunities. Elephant sanctuary, Sri Lanka.*

Source: Author, 2003.

Recently, a Brazilian ethnobotanist and agriculturalist working in eastern Amazonia produced a cartoon picture book, 'Fruits and Plants used for Life' (*Fruitiferas e Plantas Uteis na Vida Amazonica* 2003; edited by P. Shanley and G. Medina), aimed at influencing local people to value biodiversity. This was a bold move for academics to invest precious research into a publication, which might offer little professional kudos, even though it is of great practical value – a dilemma worldwide (Pye-Smith, 2003). The cartoons illustrate to farmers with little education what return they would get if they retained trees, rather than sold them to loggers. The longer-term benefits are clearly contrasted with immediate one-off gain. This blending of scientific study and local knowledge, presented in a very accessible form, promises to improve livelihoods and biodiversity protection – and might well be a strategy which could be profitably applied elsewhere in developing countries for, say, soil conservation and other 'accessible' environmental management advice.

As mentioned earlier, there are often attempts to combine biodiversity conservation with sustainable livelihoods, for example:

* *Tolerant forest management*: the extraction of products and some cropping, whilst striving to maintain as much of the original forest or other vegetation cover as possible.
* *Extractive reserves*: conservation areas where local people or other approved groups can remove products in ways that do minimal damage. The extraction may help pay for the conservation. Brazil first created extractive reserves in 1990.
* *Green tourism*: efforts are made to reduce environmental impacts and possibly use some of the profits for environmental management; the tourism is often dependent on natural features or wildlife. Green tourism includes: golf courses, which restrict pesticides and herbicides and seek to encourage wildlife; waterspout resorts, which try to protect reefs and marine life; and archaeological sites, which pay for roads and generate funds for local communities (Figure 5.2).
* *Ecotourism*: this is a stronger form of green tourism, which contributes a significant portion of profits to environmental management. Typically local

Figure 5.2 (a) *Environment and tourism. Indigenous people (Orang Asli) sell forest products, including ferns, butterflies, and orchids, at the roadside in the Cameron Highlands (Peninsular Malaysia). In many countries wildlife is for sale to tourists or is exported. Nature provides a tourist attraction and some earning opportunities, but often little or nothing is reinvested in environmental management, and controlling the trade can be difficult. Also, this trade is not a positive educational experience for the buyers. Behind the scenes collectors are probably supplying herbal medicine merchants, and extracting bamboo and rattan for furniture manufacturers.*

Source: Author, 2002.

Figure 5.2 (b) *Environment and tourism. A rope walkway strung through the forest canopy to allow fee-paying visitors a view of flora and fauna without trampling or needing forest trails cut. Some of the funds are invested in conservation and the public gets a positive educational experience. Penang Hill (Peninsular Malaysia).*

Source: Author, 2002.

Figure 5.3 *Maya pyramid (Chichén-Itzá), Yucatan, Mexico. Ecotourism development. This and other archaeological sites in the region attract large numbers of visitors, but few of the clearings cover more than a few score hectares and, apart from access roads, there is little disturbance of hundreds of square kilometres of forest.*

Source: Author, 2002.

people act as guides, and staff the accommodation and other services; the attraction is usually wildlife, scenic beauty, archaeology or tightly controlled hunting and fishing or photographic tours. An attempt is made to keep the impact of tourists and support facilities to a minimum, and educate the tourist to respect nature. Sometimes, tourists even pay to work on environmental projects (see Figure 5.3).

Community-based conservation has been spreading since the mid 1990s, with a number of aid agencies supporting initiatives; for example, the Communal Areas Management Programme for Indigenous Reserves (CAMPFIRE) in Zimbabwe (Hasler, 1999). Typically efforts are made to involve and empower locals, tap their knowledge and skills, and ensure institution building provides stable foundations for supporting conservation (Leach et al., 1997; Pimbert and Pretty, 1995).

Frequently there are excessive claims for the benefits of poorly managed ecotourism. There may also be a problem of seasonality, so that ecotourism only provides a satisfactory livelihood for part of the year. As with other forms of tourism, there is a certain amount of vulnerability to sudden changes in fashion or scares about travel caused by distant crises, which can suddenly hit profits and any environmental management which has come to depend on them (Tribe, 2000).

Conservation can also be linked to other benefits; for example, a forest reserve may be managed to lock up atmospheric carbon to counter global warming or protect a catchment providing drinking water or flows to irrigation or hydroelectric plants. In steep areas vegetation cover is vital to prevent mass movement onto land below. There is also a crucial link with commercial research and development (Swingland, 2003; Brown et al., 2002; Kate and Laird, 2002). A conservation project might even be integrated with several beneficial activities; an example in a developed country is the Great Barrier Reef World Heritage Area Strategic Development Project, which links conservation with over 60 organisations in reef-related food production, job creation, recreation, cultural heritage protection and so on (Raymond, 1996).

Developed countries benefit from funding biodiversity conservation research and monitoring. Where there are commercial pressures leading to biodiversity losses, developed countries can support efforts to seek alternatives: for example, developing substitute products. Developed countries can fund advertising to counter fashions that endanger species. Treaties and controls can help reduce the trade in endangered species. Sales controls and taxation can reduce trade in endangered species and fund conservation activities, as can tourism taxes, airport taxes and levies on traded items – provided they are not judged to be a trade barrier. The processes of globalisation and free trade agreements need to be closely monitored to ensure they do not trigger biodiversity losses.

Nowadays, people are using more products derived from areas well away from where they live, so citizens in developed countries have an impact on poor nations as a result of new trends like buying exotic fruit, vegetables and flowers. They should be made more aware of this and of their obligations to pay something toward resolving any problems their consumer habits cause. Structural adjustment packages of financial assistance for countries in economic trouble have been agreed on the condition they impose financial austerity measures. Bodies like the International Monetary Fund and World Bank have promoted structural adjustment since the 1980s, but the conditions they impose have had negative impacts on the environment of some countries. For example, farmers may have lost subsidies and turned to more exploitative land use off-farm, such as charcoal production, which damages the environment and wildlife. Another problem for biodiversity in developed and developing countries is the replacement of large swathes of natural vegetation with commercial monocultures: annual crops, tree plantations of eucalyptus or softwoods, oilpalm, rubber and so on.

One means of funding biodiversity conservation and other environmental management has been the, sometimes controversial, debt-for-nature swap (Tisdell, 1999: 63). Typically, a body like an international conservation NGO pays off some of the foreign debt for a country which otherwise might well default; in return the country undertakes to spend some of its national currency on conservation or other environmental works. In theory both the debtor nation and the environment benefit: the former saves face by not defaulting and has some control over the environmental activity, and the environment gets funding which would have been unlikely without the swap. Also the NGO has promoted environmental care. Many conservation activities are expensive to establish, demanding land purchase, infrastructure, vehicles and so on; but cost much less for ongoing management, so foreign aid and measures like debt-for-nature swaps can provide crucial start-up money, and the developing country can then find recurrent funding.

Even when satisfactory conservation areas are established, they are vulnerable to warfare, encroachment by squatters, fuelwood collectors, illicit miners, poachers and so on. Some reserves are too small or are in localities prone to environmental change and are not secure long-term. Linking a number of such reserves with corridors, which enable species to disperse, can sometimes help. Another beneficial measure is to establish buffer zones around reserves to reduce external impacts on the actual conservation area (Oldfield, 1988). Although a reserve may be large enough to maintain biodiversity indefinitely and have enough resilience against expected climatic change, there are still disasters that can strike: bushfires, the arrival of pest species, strategic demands in time of strife and pollution from distant activities. As India, China and other countries industrialise, acid pollution may well affect biodiversity in quite remote situations currently considered 'safe' and secure. Conservation efforts in New Zealand were recently endangered by one solitary 'eco-terrorist' who introduced predatory opossums to island bird reserves because for some reason he felt disgruntled.

Some countries and a number of development agencies and funding bodies are now insisting on biodiversity impact assessments for projects or policies deemed to pose a potential threat. Also, a number of environmental management systems have recently integrated biodiversity concern into their standards and prompted companies to publish biodiversity action plans outlining what they intend to do to improve things (Porter and Brownlie, 1990; Barrington, 2001).

Natural resource exploitation impacts on biodiversity

Developing countries commonly implement natural resource development projects, sometimes to generate export earnings. Sometimes there is direct mismanagement of biodiversity – forest clearance for timber, ranching or cropping; fisheries over-exploitation; extraction of wild products like rattan or ivory, wildlife trapping for the pet trade and so on. There are also indirect impacts, triggered by mineral exploitation, hydroelectric development, cement manufacture, industrial pollution, tourist disturbance and suchlike. For some of these threats to biodiversity, simple legislation and conservation efforts may bring marked results. But when biodiversity damage is a side effect of economic development, controls may be more difficult. As prospecting and survey methods improve, and as demand from developed countries increases, mineral resource exploitation is likely to remain a challenge for environmental managers. There is also growing pressure for agricultural land and urban sprawl in many countries, which impact on biodiversity.

Developers can pursue the following resource exploitation options:

- Mining of non-renewable mineral resources: the source of income is exhausted, not sustained. Impacts include noise, pollution of streams and air pollution associated with cement production, lead, copper and tin smelting.
- Production of renewable mineral resources: this is potentially sustainable and includes sea-salts, sulphur from volcanic eruptions, nitrates.
- Sustainable yields of natural commodities from well-run plantations, farmland or natural vegetation: palm oil, natural rubber, rattan, bamboo, waxes, beverages, fibres and many other materials. This can be a problem if it takes land from smallholders and drives them to marginal areas where they can damage the environment. Commodity production may take land out of food production and make countries more reliant on importing grain. It can displace and destroy biodiversity, and can be a source of pollution as a consequence of processing (see Figure 5.4).
- Unsustainable production of natural commodities: excessive hunting of species, over-collecting wild plants, clumsy ranching and plantations which degrade the land (Eden and Parry, 1996). Recently demand for herbal remedies and medicinal plants has led to over-collecting of wild biodiversity in many countries. Animals like tigers and rhino are suffering through hunting to supply traditional medicine. In Africa especially a wide variety of wild animals are being hunted to extinction to provide growing cities with 'bush meat' which is believed to have magical qualities as well as being a delicacy.

The unsustainable use of biodiversity and biodiversity 'theft' is nothing new; between the 1850s and 1880s English, Spanish, German and Dutch adventurers strove to steal seeds and plants of the best *Chinchona* species from Andean forests of Ecuador, Bolivia, Peru and Venezuela – the tree bark yielded between 1 per cent and 14 per cent quinine, which for nearly a century before chemical synthesis was the only effective treatment and prophylactic for malaria. *Chinchona* bark helped rid

Europe of the disease and greatly aided colonisation of Asia, Africa and elsewhere. Until biodiversity theft took place and the Dutch and English established plantations in the late nineteenth century the only source was from the wild and it was jealously controlled by native collectors, merchants and the governments of Bolivia, Columbia, Ecuador and Peru. The collectors failed to replant the stock they destroyed. So by the 1870s many of the best quinine-yielding species were on the verge of extinction. Those trying to steal for profit or to make cheap quinine available to the people of India could now claim they were conserving species going extinct in the wild (Honigsbaum, 2001). Natural rubber biodiversity theft, from Brazil in the early twentieth century by the British, was purely profit-driven or for strategic reasons. Once quinine and rubber had been removed the countries of origin suffered serious loss of revenue.

In the past profits have usually been taken from exploiting and damaging resources without any money put aside to rehabilitate the land afterwards – countries like Malaysia have a legacy of tin-mine craters and large areas of infertile sandy tailings (waste sediment). Some Pacific islands that were mined for

Figure 5.4 *Jarí (Brazilian Amazonia) rainforest destruction and reforestation with exotic monocultures of fast growing trees. The pulp-production factory also causes regional air pollution and river pollution, and people have settled nearby hoping to find employment. This is one of two Japanese-made paper-pulp production facilities sited in the middle of the forest. The plan was for fast-growing exotic tree plantations to supply the woodchips; in practice the plantations were largely unsuccessful and much of the supply has come from further forest destruction. A clumsy attempt at sustainable production.*

Source: Author, 1982.

phosphates lost their topsoil and now have a degraded flora and fauna; some islands were used for atomic weapons testing or as air bases; the latter in a number of cases lost biota as a consequence of rats, cats and snakes arriving.

Large mining operations and small bands of miners in various parts of the world spread diseases, hunt out game animals and pollute the environment with compounds like mercury which kill fish and other organisms even at low concentrations (Cleary, 1990; Warhurst, 1999). Papua New Guinea has serious difficulties with the waste from large copper mines polluting rivers. Aluminium production generates problems associated with large-scale open-cast mining and demands large amounts of electricity for processing the ore. Many of the developing world's large dams service mineral extraction and processing, and cause serious environmental and socio-economic problems (Barretovianna, 1992). Tropical rivers have lost fish, aquatic mammals and other organisms because dams prevent migration, regulate flows and discourage spawning, or simply cause pollution and settle sediment, reducing downstream nutrient supplies. Floodlands and estuarine ecosystems suffered because dams reduce or stop seasonal flooding – which has tremendous impact on biodiversity. The solution is for dams to periodically release flows to replicate natural flooding. However, managers are driven by short-term economic gain and hesitate because it would reduce electricity generation or irrigation storage – in a few countries, notably South Africa, artificial flooding regimes have been established. Tropical river pollution and disturbance by dams and other developments means that many potentially useful fish are lost before they can be assessed to see if they have aquaculture value, so the world is poorer. The problem is expressing the value of biodiversity in economic terms, in ways developers will heed. The cartoon book mentioned earlier in this chapter at least makes a start.

Ideally, a mining scheme should be required to store topsoil for rehabilitation and bank funds for restoration work, pollution control and compensation in the event of unwanted impacts. In time abandoned mining land can be of considerable conservation value even without restoration, especially if it is managed to encourage biodiversity. Unfortunately, countries are often keen for development projects to earn foreign exchange and provide employment and waive or weaken rehabilitation requirements. Where land is scarce, abandoned land may be grabbed by speculators and squatters, rather than be left to harbour biodiversity. Restoration can also mean planting trees that are unsuitable for wildlife conservation, but are cost-effective and aesthetically acceptable, like eucalyptus.

Environmental managers must explore practical ways of achieving 'responsible' mining and quarrying. In some cases abandoned open-cast mines and quarry excavations can be used for refuse disposal – hopefully with suitable lining to prevent leached waste polluting groundwater or streams – aquaculture and recreation as well as (or combined with) wildlife conservation. The Eden Project in Cornwall (UK) has turned a disused quarry into a tourist attraction and important conservation facility which seeks to conserve tropical and subtropical biodiversity and offer other services such as research.

Wetlands – swamps, floodlands, estuarine marshlands, mangrove forests, lagoons and so on – are being lost at a worrying rate and are some of the most valuable environmental assets available to developing countries (Roggeri, 1995).

Some of these are rich in biodiversity, and many also act as vital 'stepping stones' for migratory species. Damage to a few or even a single wetland may have serious impact on migratory species of birds and insects. The main initiative seeking to protect wetlands is the 'Ramsar Convention', the Convention on Wetlands of International Importance Especially as Waterfowl Habitat. Many countries signed this in 1971, and it maintains a permanent bureau which works closely with the World Conservation Union (IUCN) and actively monitors and seeks to support wetland conservation.

Wetlands are widely settled and some of their people have developed some of the world's best-proven sustainable agriculture strategies which could have wider application (see *chinampas,* Chapter 7). Wetlands are often drained for irrigation schemes, malaria control, real estate development and ranching. Larger wetlands like the Sudd swamps in the Sudan and the extensive marshes of Iraq have already suffered through drainage and warfare; in South America the huge Pantanal wetlands could be damaged by proposed navigation and land development projects. In Amazonia areas seasonally flooded by those rivers carrying fertile silt – *várzeas* – are increasingly developed for large-scale rice and ranching; the impact on the floodland and riverine biodiversity is considerable. Tropical forest conservation has attracted the attention of aid donors and charities; tropical wetlands and grassland savannahs also host rich biodiversity, but have attracted less support. Where exploitation is taking place environmental managers could help ensure environmental impacts are better controlled. There is a literature on wetland environmental management in Europe and North America (Turner and Jones, 1991; Turner, et al., 2003; Mitsch and Gosselink, 1993), but this needs some refocusing to meet developing country needs.

Apparently sustainable production of natural products can be less attractive than first glance suggests; plantations of crops like rubber, oil palm, timber or wood-pulp trees are often sited on common land. The enclosure of these plantations results in local biodiversity reduction and dislocation of people who may then clear and destroy flora and fauna elsewhere. There is relatively limited experience of plantation sustainability – few predate the 1950s, so it is uncertain how long they will be productive and how often trees can be renewed. There has also been little attention directed to finding ways to make plantations more able to support wildlife (Berkes, 1989).

Damage can be done by exploitative annual crop extraction; soya production is currently expanding into Amazonian forests, having already helped reduce wildlife across large swathes of less humid savannah. The spread into forest areas has been prompted by the development of varieties that can withstand humid climate and unfavourable soils and the buoyant market for non-GMO soya. For every litre of palm oil, latex rubber or methanol for motor vehicles produced, there are considerable quantities of agrochemicals used during cultivation and waste generated during processing – pollutants which can decimate riverine species. Often, export crop processing is decentralised to ensure the produce does not deteriorate before it reaches the factory; consequently, if effluent controls are not effective, pollution is likely to be widespread. In Malaysia oil palm and rubber processing polluted a large number of streams in the 1960s and 1970s.

Aquaculture is often hailed as a route to improved and sustainable food production; in practice much of the huge expansion in developing countries is not sustainable and has marked off-site impacts. Several South American, Asian and Southeast Asian countries have serious problems as a result of developing lucrative tiger prawn aquaculture. Mangrove areas are destroyed to construct ponds, water resources are exploited and returned to the environment contaminated with waste and chemical pollutants. In order to feed the prawns inshore waters may be netted for juvenile fish and plankton, which also destroys wildlife.

Less environmentally damaging prawn and fish aquaculture strategies should be possible but require changed habits and investment. Presently developers can make big profits from unsound production methods and have little incentive to improve things. The developing country government is keen to support such production for foreign exchange earnings. The environmental manager faces a challenge – how to break such established malpractices and prompt a change to better approaches. Establishing a non-polluting intensive 'closed-cycle' prawn or fish aquaculture unit is still rare, and added to this is the need to find sustainable and unproblematic sources of stock and feed.

Where natural products are being overexploited the problem can sometimes be effectively addressed by imposing controls, such as extraction quotas, wardens, export restrictions, exploitation seasons, limited numbers of extraction licenses and compulsory methods; profits from the permits and charges can be ploughed into environmental management, but often get used for other things. It is also possible to establish protected areas where some biodiversity is retained but extraction is also allowed, or develop in-vitro production or substitution with an alternative product. Natural resource production has generated considerable hindsight knowledge, so it should now be possible to plan future developments and retrofit those in place to cut unwanted impacts; the economics of production often discourage this (Epps, 1996).

There has been considerable interest in using bioregional planning to support conservation. A bioregional approach promises to be a practical way of dealing with conservation and other environmental activities at a manageable scale (Sale, 1985; Stolton and Dudley, 1999: 215–23).

Worldwide, since the 1970s, indigenous peoples have been establishing and exercising their rights over natural resources, notably in the USA, Canada, New Zealand, Brazil, Mexico and Australia. Some have made progress in improving livelihoods without untoward environmental damage. But the idea that indigenous peoples are always sympathetic to their environment, and so will not exploit it, is wishful thinking. There are unfortunately many cases of indigenous people causing environmental degradation, and sometimes central governments have found it difficult to intervene because rights have been granted. If local people do not understand and support resource development controls it will be difficult to make these work: the locals know the land and its biota and if motivated make excellent wardens to police and manage conservation or other managed areas; if they are not involved their resistance is likely to be strong. A recent development – the extractive reserve – generally seeks to involve local people to combine conservation and sustainable exploitation with local livelihood provision (Stolton and Dudley, 1999: 215–23).

The way forward for conserving much of the world's biodiversity *in situ* is through management partnerships involving individuals, communities, companies and governments (often in combination). Through these partnerships, collaboration can develop between those stakeholders wishing to exploit and those wishing to conserve.

Concluding points

- Where biodiversity conservation and livelihoods conflict there is a need to engineer solutions, which dovetail the two. Possibilities include ecotourism, extractive reserves and tolerant forest management. If local people participate in conservation it may be more likely to succeed. However, participation is not a guarantee of successful conservation – monitoring is vital.
- Tourism, NGOs, and other funding measures are increasingly paying for biodiversity conservation. Given the value many developed countries have had from biodiversity 'acquisition' from developing countries – quinine, rubber and much more – it can be argued that the North is paying too little to the South for supporting conservation efforts.
- Funds will probably be found for conserving biodiversity resources thought to have commercial potential, but efforts also need to be made to support other less potentially valuable organisms and material not seen by the public to be attractive.
- *In situ* conservation is preferable, but wherever possible conservation should spread risks and duplicate more than one reserve and *ex situ* genetic collections.

Further reading

Graves, J. and Reavey, D. (1996) *Global Environmental Change: plants, animals and communities*. Longman, Harlow. [Examines global environmental change and its impacts on plants, animals and biological communities.]

Chiras, S. and Renwick, K. (2004) *Exploitation, Conservation, Preservation: a geographic perspective on natural resources use* (4th edn.). Wiley, New York (NY). [Biodiversity from a geographical perspective.]

Shiva, V. (2000) *Tomorrow's Biodiversity*. Thames and Hudson, London. [Shiva champions the rights of peasants and plants against commerce and neo-colonialism; readable and thought-provoking, but with over 6,000 million people to feed some way has to be found to improve agriculture; this author offers little in the way of a clear alternative.]

Fowler, C. and Mooney, P. (1991) *The Threatened Gene: food, politics and the loss of genetic diversity*. Lutterworth Press, Cambridge. [This book focuses on the loss of traditional crops – Mooney has been one of the more active lobbyists on seed rights.]

Dutfield, G. (2002) *Intellectual Property Rights and Biodiversity*. Earthscan, London. [Explores the trade–environment relationship, especially the links between intellectual property rights and conservation.]

Smit, T. (2001) *Eden.* Bantam Press, London. [Story of the UK Eden Project – a conversion of an abandoned quarry into a series of biodomes. The project has some value as a conservation collection and as a research and training unit; also, like botanic gardens and zoos, it is a way of educating developed country citizens about the value of biodiversity conservation which may generate support for efforts in the South.]

Stolton, S. and Dudley, N. (eds) (1999) *Partnerships for Protection: new strategies for planning and management for protected areas.* Earthscan, London. [Draws on IUCN and WWF sources and presents excellent papers on recent conservation developments. Ranges over most ecosystems of the world and presents information on conservation and biodiversity protection management, which works with local people.]

Websites

The World Conservation Union (formerly International Union for Conservation of Nature and Natural resources) (IUCN): http://www.iucn.org/ (accessed March 2004)

The UNEP World Conservation Monitoring Centre (WCMC) maintains a database with details of the world's protected areas, budgets, information sheets, etc.: http://www.unep-wcmc.org/right.htm (accessed March 2004)

IUCN *Red List* – regularly updated list of known endangered species: http://www.redlist.org (accessed April 2004)

Convention on International Trade in Endangered Species (CITES): http://www.cites.org/ (accessed April 2004)

Botanic Garden Conservation International (BGCI) – world's largest botanic garden organisation focused on conserving wild plant material: http://www.bgci.org.uk/ (accessed April 2004)

World Wildlife Fund (WWF) – conservation of endangered species: http://www.wwwf.org/ (accessed April 2004)

World Rainforest Movement: http://www.wrm.org.uy/about/portadaEnglish.htm (accessed April 2004)

Ramsar Convention (1971 Convention on Wetlands of International Importance Especially as Waterfowl Habitat) – wetlands and wetland biodiversity protection: http://www.ramsar.org/key_iucn_agree.htm (accessed April 2004)

Eden Project – biodomes constructed in an old quarry which have been stocked with tropical plants as a tourist attraction, educational and research facility. This may help raise UK public awareness of conservation issues and offer some *ex situ* conservation: http://www.edenproject.com/ (accessed April 2004)

6 Atmospheric issues

Key chapter points

- This chapter examines how climate affects development and how human activities appear to be starting to affect climate. Consequently, there is more uncertainty as to possible future change and probably increased vulnerability. Developers must try to ensure adaptability is enhanced and vulnerability is diminished.
- Some climatic episodes seem to develop in a sufficiently predictable way that, even if their start is not accurately forecast, their progress is established and several months' warnings of adverse weather can be given. Droughts, failed monsoons and storms are likely to be better forecast.
- Transboundary pollution is a relatively new challenge. Acid deposition, particulate air pollution, atomic fallout, ozone scavenging, release of genetically modified organisms, greenhouse gas emissions and many other atmospheric contaminants must all be carefully monitored. National governments, international bodies, experts and law are still adapting to cope with transboundary problems so agreements and controls are not easy to achieve.

Between the 1860s and 1930s a number of environmental determinists explored links between human development and climate, only to be dismissed as simplistic and in error (Glantz, 2002; Barrow, 2003: 1–31). A few historians who have conducted thorough research that suggested clear human–climate linkages have escaped being dismissed as determinists (e.g. Le Roi Ladurie, 1972). During the last 30 years or so there has been increasing interest in natural climate fluctuations and their possible impact. Severe El Niño/Southern Oscillation (ENSO) conditions have proved costly for several countries and it has become clear from palaeoecology and historical records that conditions fluctuate enough to have serious effects and that there is at least quasi-periodicity. Improved modelling holds out the possibility that ENSO events (both El Niño and La Niña episodes) might be forecast. Even if that proves impractical, enough is known about how these episodes unfold to make it possible to warn some countries of coming changes 12 months or more in advance once conditions start to shift in key locations. Environmental managers should thus have better warning of extreme events like drought, failed monsoons and heavy precipitation and storm risk. This is something colonial administrators dreamed of – in India efforts were made to try and develop early warning for poor monsoons (Davis, 2001).

Added to natural fluctuations are climatic changes that appear at least part-caused by human activities. Atmospheric pollution has probably altered natural climate change: both ongoing shifts and the behaviour of episodes like ENSOs. The consequences are far from clearly predicted, but are likely to include global warming, perhaps sudden regional climatic shifts to less warm conditions, altered ENSO timing and severity, changed severe weather patterns, and shifts in rainfall and drought occurrence. Environmental management must be prepared for unexpected changes and should lobby for development to make allowances for uncertainty. Conservation areas should have the capacity to adapt to climatic shifts or flooding; large development projects, infrastructure like drainage and – crucially – food supply should be designed to be adaptive.

Regional, national and transboundary pollution

Transboundary pollution is a relatively new phenomenon. From Roman times materials like lead have spread globally; however, since the late 1940s, atomic fallout, organochlorine and other pesticides and compounds like polychlorinated biphenyl (PCBs) have been released and have posed problems at an international scale. Increasingly since the 1940s, people have been exposed to global background pollution. Starting in the 1930s, new compounds have been created which between the 1970s and 1990s were proven to cause stratospheric ozone damage. The impact of the last 200 years or so of carbon dioxide and methane emissions was only accepted to be a serious issue in the mid 1980s. While sulphur compounds and other industry-, power generation- and traffic-related pollution were seen to be local or regional problems even in medieval times, it was only after the 1970s that acid deposition and other air pollution was perceived to be a serious transboundary threat. Developing countries have become involved a little more recently as they developed industry, adopted new compounds, cleared forests and developed agriculture.

Acid deposition

Pollutants generated by industry, traffic, domestic heating and so forth drift as 'dry' particles or in cloud droplets, mist, drizzle, rain or snow. The 'cocktail of pollutants' includes many unpleasant and harmful materials, soot and often a good deal of sulphur compounds which react with atmospheric moisture to cause acid droplets. Dry and 'wet' material can be deposited as acid deposition. This may be relatively steady or episodic, the latter more likely where there are shifting winds or downpours.

Acid deposition is generally recognised when precipitation has a pH of less than 5.1; normally, it would be slightly acid at around 5.6. It is not unusual for acid deposition to have a pH of 3.0 or even as low as 2.3, similar to lemon juice or battery acid.

Vegetation, including crops, can intercept dust and contaminated mists blowing past and accumulate considerable amounts of pollution before it is suddenly washed off by heavier rain or released after leaf-fall (see Glossary: chemical and biological time bombs). The composition of this pollution varies but all cases share the characteristic of being acidic – hence the popular expression 'acid rain'. Prevailing winds, aspect and exposure, surface characteristics, the distribution of pollution sources and many other factors cause marked local variation in the pattern and strength of acid deposition. The reaction of organisms, even individual plants of the same species, may differ in response to a given level of acid deposition. The problem is generally most marked 'downwind' from sources and may cross national or federal boundaries (McCormick, 1988; Wellburn, 1988).

The pollution effects can be subtle and sometimes cumulative and manifest in ways that are difficult to trace back to source, so there are challenges in monitoring and control. It is possible to map various soil and vegetation characteristics to generate maps of vulnerability and to use natural indicator species and instruments to measure the impact of acid deposition.

Elevated acidity has the following impacts:

- It damages man-made structures: paint and plastics deteriorate, metals corrode faster, buildings – including historical sites – weather badly.
- It alters the metabolism of organisms and affects biogeochemical processes in water and the soil.
- It injures plants and other organisms: it can burn vegetation, and may alter the pH of soils and water bodies; even slight change may have serious impacts on organisms which prefer non-acidic conditions. Changes in soil structure, altered populations of micro-organisms and the mobilisation of plant nutrients and toxic compounds like phosphorus, heavy metals and aluminium may be devastating but difficult to measure and track.

Soils and water bodies vary in their ability to counter acid deposition; alkaline environments are likely to buffer the effects more. In Europe and North America where the problem was first recognised in the 1970s, it was marked by tree disease, die-back and death; changes in upland vegetation cover; and serious damage to the biota of many streams, lakes and marshland. Soils which are acidified may release toxic compounds, and some groundwater and surface supplies have become dangerously contaminated by heavy metals as a consequence. Worldwide, upland forests are especially vulnerable because the vegetation is likely to be under stress, growing on thin soils and exposed to passing polluted mists. Acid soil areas are also vulnerable because they have little capacity to buffer (counter) acidification.

In the tropics, soils at low altitude are often well weathered, already acidic and are commonly rich in aluminium or iron compounds. Those metals are easily mobilised by acidification, leading to reduced soil fertility, which makes agriculture more costly, difficult to sustain and labour-demanding. The response of adding chemical fertiliser or lime leads to more agrochemical runoff. Acid deposition causes depressed crop productivity; damage to forestry and plantation crops; damaged fisheries; and unwanted impacts on biodiversity. Even well-managed and little-disturbed conservation areas may be vulnerable to acid deposition from distant sources. Vegetation damage and kill-of organisms may be gradual or sudden, and seems to result from a diversity of apparent causes, such as pests, diseases and extreme climatic conditions, but ultimately are a consequence of altered environmental acidity. Acid pollution damages infrastructure, increasing maintenance costs and can injure animals and humans, causing respiratory problems, eye damage and other problems.

Where soils have accumulated pollutants like pesticides over a period of time there is a risk that acid deposition may alter soil chemistry and cause a sudden damaging release. Relating overlaying maps of acid soils, biodiversity conservation sites and vulnerable ecosystems to regions where industrialisation is increasing makes it possible to predict future problems areas (see Figure 6.1) (Rodhe and Herrera, 1988; Rodhe et al., 1992). There may be positive impacts: disease-carrying snails may be hindered from breeding if water bodies become acidic; certain crop diseases may be inhibited; and some grasses and cereals may grow better when there are limited levels of acid deposition. Tracking changes is difficult, and proving the acid pollution–effect relationship may take time and raise challenges for legislation and international controls. The problem is costly to treat after pollutants have been

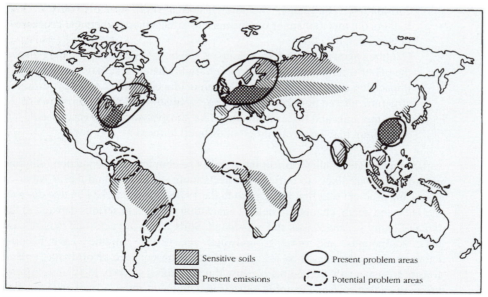

Figure 6.1 *Acid deposition threat: regions that presently have a problem, areas of sensitive soil, and areas likely to suffer increased pollution.*

Source: Based on a schematic map in Rodhe and Herrera (1988), fig. 1.8, p. 27.

emitted; some regions have resorted to using alkaline compounds, especially lime, to treat soils and water bodies. Such treatments are seldom wholly successful, so ideally, acid emissions should be prevented.

The problem is especially related to the combustion of sulphur-rich coal, oil and other fuels, but similar problems may also be triggered by the use of chemical fertilisers or irrigation which cause some soils to acidify; it may also happen naturally when acidic, sulphurous ash and gasses are released by some volcanic eruptions, perhaps as a result of huge methane emissions from ocean sediments, when sea spray contaminates precipitation, as a consequence of transpiration by forests, and through a few other processes.

India and China are increasingly likely to make use of their sulphurous coal as they develop – without pollution control, acidification will be a problem. Technology for sulphur reduction is available and effective; even in the 1920s, Battersea Power Station in the UK could scrub up to 90 per cent of its emissions as they were released, and technology has greatly improved. Another possible solution is chemical desulphurisation, washing or biodesulphurisation to reduce the capacity of coal to pollute before it is burnt. Pollution reduction is possible and relatively straightforward, but countries may be reluctant to bear the costs of doing so. Consequently all nations should be involved.

Acid deposition has become a growing problem in developing countries (Park, 1987: xii; Rodhe, 1989; Rodhe and Herrera, 1988; Rodhe et al., 1992). Recent studies suggest a number of Asian cities will have acidic sulphur air pollution exceeding World Health Organisation (WHO) safety levels by 2020. In the last 25 years acidic sulphur air pollution has more than quadrupled across India and South East Asia (Ananthaswamy, 2003). In addition to local and regional acid deposition, there is a problem at the global level with pollution apparent at remote sites like the poles, tundras and mid-ocean monitoring stations. Globally, there has been a roughly three-fold increase in acid deposition over the last 100 years or so, and allowing for normal natural sources around half probably results from human activity. Where wildlife, and possibly livestock, feed on slow-growing vegetation, like lichens or grasses in high altitude or polar environments, accumulated acid deposition can cause them to be contaminated with heavy metals and other harmful pollutants. Tundra and forest plants are also affected and this could affect food chains. The impact on oceans is not established. In 1993 the World Bank became sufficiently concerned to fund an Asian Acid Rain Monitoring Project.

There can be large natural emissions of sulphur compounds, some sufficient to cool world temperatures for months or years. Palaeoecologists have identified what they think are such events in the past: there are signs this may have happened in 1816, which saw cool summers in Europe and elsewhere. Human activity could raise background levels so that natural outbursts may have more impact. The possibility is that such a cooling event, even if it lasted only a year or two, could catastrophically hit world food production. Perhaps these risks could be researched and made more apparent to the public to encourage payment for controls.

Critical load – a useful benchmark concept – is the amount of acidity an environment can absorb on an ongoing basis without significant damage. Using such concepts the environmental manager can model and monitor acid deposition, work toward controls and encourage alternative cleaner sources of energy. The solution will largely depend on developed countries providing aid and technological support; one funding source is the much-criticised Global Environmental Facility (GEF), administered by the World Bank (Young, 2002).

Airborne gases, particles and aerosols

There are a number of pollutants that are commonly present with the sulphur compounds just discussed. For some years there has been concern about 'haze' and 'smog', a mix of fine suspended particulate matter, soot, aerosols and gases especially sulphur dioxide. The composition varies, reflecting the sources involved, weather conditions and how long it has been exposed to the atmosphere and sunlight – many regions now have regular occurrences.

South East Asia, especially in 1997 and 2002, had difficulties with 'brown haze' derived from biomass burning, plus industrial combustion of coal and other hydrocarbon fuels and motor vehicle emissions. The problem was largely due to Indonesian farmers using fire to clear forest and grasslands. The problem seems to have resulted when the forest clearing seasons coincided with dry conditions caused by ENSO events. The brown haze blew across Indonesia, Malaysia, Thailand and Singapore and was serious enough to cause widespread respiratory problems, wildlife and crop damage, transport disruption and ruined tourism. Similar pollution events are increasing in China, India, Thailand, Zambia, Amazonia and elsewhere. There are fears that the increasingly widespread haze and the cloud cover is blocking out as much as 16 per cent of solar radiation in affected regions. If unchecked, regional cooling might alter crop yields and water supplies. Air pollution, including particulate matter, acid deposition and troposphere ozone, poses a serious and underestimated threat to food production in many developing countries (Marshall et al., 1997). In polar and highland areas soot particles deposited from airborne pollution may darken snow and alter natural melting patterns, which may affect water supplies and regional climates at lower latitudes. There is also a chance that these hazes and smogs may have subtle impacts: for example, triggering genetic defects. The problem is unlikely to diminish unless land clearances with fire can be halted and developing countries somehow reduce hydrocarbon emissions as they industrialise. Natural forest fires are more likely when forests and grasslands have been disturbed.

Radioactive pollutants

A number of developing countries have civil nuclear installations, and several have weapons facilities. Some send high-level waste to developed countries for reprocessing; others are disposing of waste themselves and releases to the atmosphere or oceans are not impossible in-country or en route for reprocessing. There are developing countries running reactors of less-than-ideal engineering standards – worrying in the light of Chernobyl experiences. Those who were reliant on Soviet technical assistance may pose a particular risk because of cutbacks and problems with spare parts. Where there is a risk of earthquake, tsunami and a warm climate which causes concrete to deteriorate fast, the chance of a failed containment vessel is greater.

There may be a temptation for a poor country to offer sites to richer nations for radioactive waste repositories in return for foreign exchange. This can be a complex issue, especially where indigenous peoples with a degree of autonomy negotiate with countries other than their own. Very radioactive waste and large quantities of less-active waste are likely to need constant cooling, high security protection and skilled maintenance, perhaps for centuries, to avoid escapes. Hospitals and industry are increasingly using isotopes and disposal presents a problem; there have been a number of tragic accidents, one in 1987 in Brazil where children were fatally irradiated and contaminated their families and friends after playing with a caesium-137 source from medical equipment left unburied on a rubbish tip. A similar tragedy occurred in Uganda in 2002 as a result of a discarded cobalt source. Another problem with poorly managed civil and military radioactive waste is that it may fall into the hands of terrorists or 'rogue states' and be used for weapons manufacture. The production of an atomic bomb from such sources is difficult; however, a 'dirty bomb' is relatively easy. Such a weapon would make use of any highly radioactive source and an explosive charge to scatter contaminated debris over a major commercial centre, strategic site or populated area. The heath risks, fear and clean-up problems raised make these a formidable terror weapon. Given that many terror groups could assemble such a weapon, and that even a pea-sized piece of highly active waste may be enough, controlling the threat is difficult. The former Soviet Union admits considerable quantities of material like caesium-137 are unaccounted for, and commercial sources are constantly discarding waste worldwide. Monitoring is probably going to be expensive and will need to involve an international body and liaison with various intelligence agencies.

Fears have been voiced that a new generation of weapons may soon appear which have the power of low-yield atom bombs, but can be made much smaller; they might also be much cheaper, and also they would remain outside present weapons control agreements because they are neither true fission or fusion devices. The possibility is that these will be much easier to manufacture than present atomic weapons and easier to conceal, transport and deliver to a target.

Global greenhouse gas emissions

'Greenhouse gas' is a term applied to various gases emitted as a consequence of human activity which are believed to 'force' global climate change; the assumption is that this will be warming. Developing countries contribute to global atmospheric carbon, and increasingly emit more powerful greenhouse gases like methane. There has been much debate in the literature and extended international negotiations over the amounts each nation has generated and continues to generate and how to tackle the problem. At times this has been heated, with accusations that rich countries are 'rigging' the data to downplay their responsibility and elevate that of poor nations. Whether or not that is true, a number of developing countries have cleared considerable amounts of forest and grassland, and most have disturbed their soils and released carbon that had been sequestered there. There is also considerable irrigation development, use of chemical fertilisers and expansion of large ruminant herds – all of which emit methane. Added to that are the emissions from industrial expansion in developing countries. Recent estimates suggest that by 2010 about 50 per cent of global warming will be caused by developing countries, especially India and China.

So, all nations are involved in causing global environmental change. Any attempt at control needs to impartially calculate and agree national carbon and other forcing gas emissions and then negotiate a globally acceptable strategy to deal with the problem. Practical agreement and enforcement require much more improvement of global change-related legislation (Churchill and Freestone, 1991). Unfortunately, agreement has not been adequate, with the USA and other major greenhouse gas-emitting countries refusing to sign the Kyoto Protocol (see Box 6.1). Part of the hesitation is fear that signing the Protocol will handicap economic development and cause unemployment.

Climate change predictions draw upon global climate models (GCMs), which are far from accurate – warming may be possible, but it is far from certain (Philander, 1998), and there might be problems that are quite different to those predicted. What might be the impacts of warming weakening Atlantic thermohaline flows and suddenly reducing the Gulf Stream? Such a development is not impossible and a chilled Western Europe might require vast quantities of food aid; world stability would be threatened (Steffen et al., 2004).

The impacts of global environmental change on developing countries may differ considerably from that on developed. The latter are undertaking much of the efforts

---

Box 6.1

Kyoto Protocol

The Protocol grew out of the UN Framework Convention on Climatic Change, which was launched in 1992 at the Rio Earth Summit to fight dangerous climate change. The parties involved have met regularly since to negotiate. The EU is the main promoter of the Protocol. In 2003 the USA refused to ratify the protocol and very few signatories are currently on target to achieve the goals set by the 171 signatories in 1997. Russia has also declined to ratify the agreement, as did Australia in March 2004. The mechanism likely to be used to cut emissions, if there is agreement, is the tradable emissions quota (TEQ). A country exceeding its TEQ allowance can buy spare quotas from one that has surplus. If not it will be penalised. The Protocol is likely to include provisions for crediting countries that avoid releasing sequestered carbon by preventing deforestation

By early 2004 most countries which had signed were unlikely to meet their treaty obligations, following a history of disagreements. Even if the Protocol is agreed by all and becomes legally enforceable, it will only run until 2012. The Protocol is seeking marked cuts in emissions, and targets will probably be based on national populations. Measures will seek 'contraction and convergence': reduced emission, leading ultimately to similar emissions restrictions for all. The ideal would be that every citizen in the world would have the same right and quota to pollute. But this is unlikely to be reached before 2050 at best. Some countries are already trying to work toward cuts of 60 per cent or more in emissions. In the short term there will be countries that find the agreement unfair because new industrial nations will have lower targets, if any.

Source: Author

to predict future scenarios and the knowledge might give them advantages. Increased atmospheric carbon dioxide content and associated climate changes may affect tropical crops and weeds in a more negative way than those of developed, higher-latitude countries. Poor nations have less to spend on responses and in some cases have huge numbers of vulnerable people. The likelihood is that richer countries and business will be better prepared and may exploit the situation for their own survival or profit (Whyte, 1996; Martens, 1998).

Already vulnerable developing countries have come together to try to protect their interests in the face of global environmental change; for example, the Association of Small Island States (AOSIS) has been vociferous at international meetings. Many of these islands have much to lose if there is even slight sea-level rise through global warming. And the same is true for low-lying regions like Bangladesh and the Nile Delta. There are various possible responses to the threat of global environmental change:

- Agree to command-and-control principles to reduce further emissions;
- Rely on market-based instruments for the same ends (Endres and Finus, 2002);
- Adapt by altering livelihood strategies as climate changes, by retreating from land vulnerable to sea-level rise;
- Mitigate global warming effects – this is more of a challenge than the previous three.

It may be possible to enhance carbon sequestration, but some other proposed responses could be dangerous and might cause even greater environmental problems.

Rules and regulations are sometimes evaded, so market-based instruments may be promising. However, at present not even the rich nations can reach agreement on global environmental change controls (Patterson, 1996; Soroos,1997; Oberthür et al., 1999; Gupta, 2001). Some countries are actively disinterested, waiting to see what happens, and hoping that others will adopt emission controls and reduce the need for them to act and spend – these are not just developing nations.

There is much uncertainty over global environmental change; modelling is not accurate, there is the ongoing risk of unexpected shifts, and Nature could throw up other threats. Planners and managers should maintain vigilance for *any threat*. Ideally, all developed and developing countries should be striving for mutual support, seeking to reduce vulnerability, and improving adaptability to as wide a range of threats as possible. Focusing mainly on global warming is unwise.

Ozone problems

Natural levels of ozone in the atmosphere have been in a steady state throughout historical times. However, it is likely that solar radiation fluctuations, volcanic eruptions and geomagnetic variation might affect stratospheric ozone levels from time to time, sometimes in a minor way and occasionally perhaps catastrophically.

Over the last half-century or so there have been changes caused by human activity, with the following results:

- Stratospheric (high altitude) ozone, which plays a crucial role in helping shield the Earth from ultraviolet radiation, has been depleted.
- Ozone, which is a reactive pollutant that damages organisms and infrastructure, has increased in the troposphere (lower atmosphere).
- Ozone is also an effective greenhouse gas and increases or decreases may have climatic impacts.

Stratospheric ozone scavenging has taken place worldwide since the 1930s, resulting in a reduction of roughly 8 per cent; however, losses have been especially marked at the poles and higher latitudes. This has resulted since the late 1980s in Antarctic and Arctic 'ozone holes'. Given the threat of occasional natural thinning, any human-caused reductions are unwelcome. The thinning increases ultraviolet radiation penetration of the atmosphere, which can damage wildlife, crops and human health. Tropospheric ozone increases do not reach the stratosphere and compensate for losses there.

The gas is produced when partially burnt and unburned airborne hydrocarbons are struck by ultraviolet radiation. The source of the hydrocarbons is the incomplete

Figure 6.2 *Urban transport, South Asia. Autotaxis (tuc-tucs): in many South and Southeast Asian cities these offer cheap, flexible public transport ideal for crowded roads. They are not, however, pollution-free.*

Source: Author, 2003.

combustion of vehicle, industrial and domestic fuel. In calm conditions, tropospheric ozone can be produced in areas quite distant from sources of pollution in quantities sufficient to have significant impacts (see Figures 6.2 and 6.3).

Overall, there has been good progress in understanding causes and consequences, and in negotiating international agreements for control of stratospheric ozone depletion. Initially the threat was seen to lie with supersonic aircraft and then propellants in domestic spray goods. Given that the causes were luxury products with identifiable manufacturers, restrictions were relatively easy, and largely achieved by the early 1980s. Supersonic travel failed to materialise enough to cause problems; however, chlorofluorocarbons (CFCs) and related compounds were widely used from the 1930s for refrigeration, air-conditioning, industrial cleaning and pest fumigation. CFCs were not seen to be a serious threat to stratospheric ozone after 1987, but once the problem was proven alternative compounds have been developed and promoted.

Agreements on the use of CFC replacements between developed and developing countries have progressed well, which gives some cause for optimism over other transnational negotiations (Rowlands, 1995). The Montreal Protocol was signed in 1987 and toughened in 1988 and 1990. It places limitations on the use of damaging CFCs, seeks gradual reduction in use and the adoption of alternatives. While controls have been negotiated and alternatives seem to work, there is unlikely to be

Figure 6.3 *City traffic pollution, Mexico City. Mexico City lies at over 2,500 metres altitude in a sheltered basin: the result is an infamous VOC-rich smog.*

Source: Author, 2001.

any significant fall in stratospheric ozone before 2050. This is because ozone-scavenging compounds are slow to degrade and there will continue to be escapes of CFCs from discarded refrigeration equipment. There is also some smuggling and illicit use of banned CFCs which are cheaper and thus attractive to unscrupulous consumers. Emission controls on vehicles can be effective in combating tropospheric ozone and other pollutants, but catalysers cost a lot and need regular replacement during the life of a vehicle. New cars may have them, but owners are unlikely to be able to afford to maintain them a few years later. That is particularly likely in developing countries.

Countries most affected by stratospheric ozone losses are those at high latitude where there are ozone layer 'holes' and thinning, and those at lower latitude with sunnier conditions and where sunlight strikes through slightly less depth of atmosphere; those with territory at high altitude, such as Andean countries, highland Mexico, highland Africa and Himalayan areas are likely to suffer most.

The effects of stratospheric ozone thinning on wildlife, crops, livestock and human health are not established. Organisms can suffer genetic damage, possibly have eyes damaged and suffer increased rates of cancer. Immunity to disease might be affected and balances between species might be disrupted. Fears have been voiced that oceanic plankton could be affected enough to affect fisheries productivity or even trigger serious global environmental change. Raised ultraviolet radiation levels and troposphere ozone both damage crops, and legume crops seem especially vulnerable to the latter – developing nations like Brazil have invested a great deal in soya and might suffer depressed yields. Some provision has been made for international funding to assist poor nations to adopt technology and policies which are less damaging to stratospheric ozone, notably the development of the Global Environmental Facility (GEF).

Volatile organic compounds

Where there are inadequate funds to support the widespread provision of mass urban transport and control vehicle exhausts, urban areas and the land for hundreds of kilometres downwind will suffer air pollution, including troposphere ozone, acid deposition and volatile organic compounds (VOCs). VOCs can cause a range of health problems and damage wildlife. Vehicle emission controls can help reduce VOCs, if motorists can afford it.

Heavy metals

Heavy metals tend to accumulate in food and water, but some become airborne as particles. The sources include vehicle exhausts and dust from clutch-linings, brake pads and tyre wear. Developing countries in particular have often retained lead anti-knock compounds in petrol. Until lead-free fuel is adopted worldwide, lead pollution in cities and close to roads will remain a threat, especially to the health of children. Where heavy metals are being mined and smelted there is a local and regional problem, especially downwind, and where they pollute rivers or lakes or the sea. More and more developing countries have heavy metal polluted industrial areas: for example, the Zambian Copper Belt. Within towns and cities of the developing countries it is not uncommon for there to be scattered workshops which are difficult to monitor, and which can pollute the air with a wide range of compounds, often including heavy metals.

Concluding points

- Once pollutants get airborne wind systems can widely disperse them. The Chernobyl disaster showed how easily and how seriously problems could affect distant countries. Some global pollution may not be so suddenly and obviously threatening; stratospheric ozone thinning was not spotted for nearly 50 years after industry started to use 'safe' compounds. Once a problem is spotted it can be a slow struggle to prove it and convince governments, businesses and peoples that action must be taken and costs borne to counter the threat. Environmental managers have to liaise with lawyers, public relations advisers, NGOs, media and whatever to support efforts to get controls. They must also be right: any false alarm or 'crying wolf' makes future threats much more difficult to deal with. International law and the outlook of national negotiators are still getting used to solving transboundary issues and global problems.
- Some transboundary pollution poses a global threat to human health, food supplies, biodiversity and environmental stability.
- Global environmental change induced by pollution is difficult to model. Predictions are not accurate or reliable, and people should not allow a false sense of security to develop. The expectation is that there will probably be gradual gentle warming, and that it will not be too disruptive. The reality is that outcomes of current changes are uncertain.
- Planners and managers must be vigilant for any threat, not just global warming, and should seek to reduce human vulnerability and increase adaptability to as wide a range of disruptions as possible. An excessive focus on global warming must not blinker the world to other dangers.

Further reading

Rowlands, I.H. (1995) *The Politics of Global Atmospheric Change.* Manchester University Press, Manchester. [Reviews the political issues associated with global atmospheric change.]

Intergovernmental Panel on Climate Change (1991) *Climate Change: the IPCC Scientific Assessment.* Cambridge University Press, Cambridge. [Libraries may have this authoritative and understandable report on what is thought likely to happen in the next half century or so. The website below has information on more recent IPCC assessments.]

Godrej, D. (2002) *The No-nonsense Guide to Climate Change.* Verso, London. [Entertaining, but not always objective, introduction to the subject.]

Houghton, J. (1997) *Global Warming: the complete briefing.* Cambridge University Press, Cambridge. [Good introductory text.]

Leggett, J. (ed.)(1990) *Global Warming: the Greenpeace Report.* Oxford University Press, Oxford. [Sound introduction prepared by an NGO but getting dated.]

Parry, M. (1990) *Climate Change and World Agriculture.* Earthscan, London. [This is a popular précis version of a large two-volume publication that seeks to predict the impacts on the agriculture of developed and developing countries.]

Ince, M. (1991) *The Rising Seas.* Earthscan, London. [Readable introduction to possible sea-level changes associated with global warming.]

Websites

Intergovernmental Panel on Climate Change (IPCC) – established by the World Meteorological Organisation (WMO) and the United Nations Environment Programme (UNEP) to assess information relevant to understanding climate change and its impacts, and options for adaptation and mitigation: http://www.ipcc.ch/ (accessed March 2004)

Latest ideas and research in changing climate: http://www.changingclimate.org (accessed April 2004)

7 Urban environments and industrial pollution issues

Key chapter points

- This chapter examines urban environmental management and non-rural pollution issues.
- Developing country urban environments are diverse and sometimes different from those in developed countries. During the last decade urban populations have expanded to include more than half of all people. Much of that expansion has been in developing countries, where some cities are now huge.
- Urban problems tend to result from rapid population growth, poverty and poor governance. City impacts are felt way beyond urban areas. Some developing country urban problems, like diseases, could spread to affect developed countries; it is therefore in the interests of citizens of the North to aid improvements in the South.
- Pollution caused by cities and industry in developing countries has got worse. This chapter looks at causes, ways to seek reduction and problems like the export of pollution hazards and pollution accidents.

Urban environments

There is no clear, simple, universally accepted definition of 'city'; indeed, in some developing countries urban areas are distinctly different from those of richer countries. Single-storey shanty towns in the cities of developing countries may not have especially dense populations, although inner-city tenement blocks might. Cities in developing countries often have large amounts of unoccupied land and few of the services familiar in metropolitan areas of richer countries. There is great diversity; along with established city areas some countries also have large refugee camps which display many of the characteristics of urban areas. Such camps can form suddenly and impose severe demands upon water, food and fuel supplies in a region, sometimes for decades; they also lack taxpayers to support services. The people of such camps have often lost possessions, and are traumatised and desperate, presenting a challenge for those seeking to improve things.

Cities may have existed for centuries, they may have sprung up in response to colonial activities and new communications networks, and in some cases are recent planned creations where little existed before. For several years now the world has been more urban than rural, in terms of population numbers, and much of the future population growth in developing countries will be in cities (see Box 7.1). Often, but not in all cases, developing country cities have grown rapidly in recent decades. Rapid, and largely unplanned, urban growth poses serious problems for environmental management (Devas and Rakodi, 1993). Here and there it is possible to find cities which have made progress in improving urban environmental management, but most of these are relatively affluent city states or have flourishing economies based on petroleum or tourism. Singapore is one such example, having made impressive progress since the 1970s in spite of limited space and industrial growth (Khoo, 1991).

There is a growing interest in Third World urban ecology (Ward, 1990; Schell et al., 1993). Some of the literature contains over-generalisation and there are commonly repeated fallacies. The following points seek to objectively list developing country urban environment characteristics:

- Cities can provide healthy and stimulating environments for their dwellers, and can generate funds and services to aid rural areas.
- Rapid population growth is not always a cause of urban problems; there are cities which have undergone rapid demographic expansion with limited ill effects. Rapid growth coupled with poor governance is problematic.

Box 7.1

The demographics of developing country urban areas

Between 1950 and 1990, the urban population of Africa, Asia and Latin America grew from about 286 million to over 1.5 billion; by 2000 it had passed 2.25 billion. An estimated 80 per cent of world population growth between 1990 and 2000 took place in urban areas.

Between 1950 and 1990, the urban population of Africa grew from 14.5 per cent of the total to 40.7 per cent; Latin America experienced growth from 41.5 per cent to 76.4 per cent; Asia, excluding Japan, saw growth from 14.2 per cent to 41.5 per cent.

A lot of people are living in smaller urban concentrations of roughly 250,000 to 1 million. Cities in developing countries often have quite low densities of occupation.

In 1950 only 10 of the world's cities exceeded a population of 1 million (the number needed then to constitute a megacity); in 1990 there were 171, which housed roughly one-third of the global population.

In 1990 nine developing country cities exceeded a population of 10 million, which is what today constitutes a megacity.

Source: Several including Drakakis-Smith (1987); Hardoy et al. (1992: 29–31).

- While there are many very large cities, and most of these have appeared quite rapidly, much of the urban environment in developing countries consists of far smaller settlements.
- Mega-cities are not necessarily prone to problems, but if problems do appear considerable resources may be required to solve them.
- There is a huge diversity of urban environments: cities can vary a lot and within a city there are often marked differences, notably between rich and poor districts.
- Statistics are incomplete and inaccurate, consequently there is too much generalisation and false assumptions about problems. Problems vary from city to city and within a single city.
- Ways in which developed countries have responded to rapid population growth and urban environmental problems in the past may not be appropriate for the future.

The environmental manager's goals are to ensure the city gives its inhabitants a sustainable healthy environment and employment, and that doing this does not unduly impact on other, extra-urban, areas. There are advantages in a city situation: in particular, the concentration of people gives economies of scale for provision of services; it is also easier to monitor and enforce environmental standards when activities are less scattered, and there are likely to be opportunities to dovetail activities: e.g. by-products can be passed to subsidiary industries. There has been a suggestion that city development should be 'de-linked' as much as possible from the rest of the world. This would involve encouraging large cities rather than smaller settlements, pursuing high-tech food production close by, and allowing remote areas to return to nature. Whether that could be a viable strategy remains to be seen. Drastic changes like de-linking are unlikely and continuation of established patterns of city management are what environmental managers will work with. The message, which comes across from various sources, is that successful cities depend upon good governance.

Planners have tried to foresee and control city problems with mixed results; few would hail Brasilia, constructed from scratch since the 1960s, as an undisputed modernist success, with its activities rigorously zoned into specialist districts and street plans drawn up with the car and modern lifestyles in mind. Yet Curitiba, another Brazilian city, one comparatively 'unplanned', has made impressive progress in coping with its environmental and socio-economic problems in spite of limited funding (Rabinovich, 1992). Curitiba has recently developed public transport strategies which rich cities like London could learn much from.

In general, the amount of pollution in cities in developing countries is less on a per capita basis than it is in most developed country cities. Poor people consume less and much of what they discard is organic material, which can be composted; in general their refuse less toxic than in rich countries. The poor also use far less heating and air conditioning, although they often depend on fuel wood and charcoal, which may cause local air pollution problems and serious damage in source areas. For example, Dhaka (Bangladesh) in 1992 emitted roughly 25 times less carbon per capita than Los Angeles or Chicago (Hardoy et al., 1992: 187). For this reason international agreements on carbon emissions, if based on a per capita consumption, might be seen to favour developing countries. Hardoy et al. (1992: 17) were at pains to stress that speed of urban growth and size of settlement are not necessarily problematic, but that growth with poor administration and planning is. When that happens, services are not adapted and expanded to keep pace.

Worldwide, urban areas markedly modify runoff, making it more erratic, usually polluted, with more extreme flows, and with the discharge channelled. Cities pollute the air flowing past, and the albedo of the built-up area may cause a 'heat island effect', causing warmer cool season and hotter warm season temperatures within and around them (Douglas, 1983; Harris, 1990; McGranaham, 1991). The supply of inputs for cities – water, energy and food – affects a considerable area, which can be measured as an ecofootprint, and may even be felt far overseas. The concept of an ecological footprint was developed by Rees (1992) to evaluate the impacts a city has on its surroundings and is now widely used.

Until quite recently, illegal settlement of urban and peri-urban land usually prompted the authorities to respond with force, evicting people, bulldozing shanty settlements and harassing the squatters to drive them elsewhere. But since the 1980s the sheer numbers involved, plus the growth of civil rights and media interest, has reduced such treatment. Nowadays, city authorities cannot ignore their slums and squatter settlements. Growing numbers of enlightened authorities have started to try to provide improved water supplies, waste disposal and public transport. Longer-established squatter settlements usually make some effort as communities to improve their services and standards of housing, and to reduce disaster vulnerability (Maskrey, 1989; 1990; Aldrich and Sandhu, 1995). Illegal settlers are often a large proportion of total city populations in developing countries; for example, in Recife (Brazil) they account for over half.

Good city governance should include strong environmental management and adopt a forward-looking approach:

- developing adequate services
- passing and enforcing appropriate legislation
- implementing protective measures.

There is also a need for city administrators and planners to look at developing country needs, rather than adopt approaches suited to rich communities and non-tropical environments. Quite often cities were established for reasons that are no longer valid; what made sense for colonial administrators or with respect to past rail or river transport may no longer hold. Environmental improvements are likely to be part of urban rethinking which finds new revenue and generates jobs. Improving services and the environment of developing country cities also demands appropriate ways of generating funds and, wherever possible, recovering costs.

Developing country cities usually have storm drains – sewers, generally open channels, which easily block and overflow. Ill-drained areas become contaminated and support the breeding of mosquitoes and other unwanted pest organisms, wells become contaminated and dwellings damp. There are many other problems: crowding, poor housing, lack of street lighting, unsurfaced paths, open gullies, uncleared sewage and refuse which harbour vermin and the use of dangerous cooking and lighting methods – particularly kerosene stoves – in flammable shacks. All this, plus a poor diet, combines to damage human well being. Those living in such districts are at risk from:

- Diarrhoeal diseases: a major killer, especially of children – for example, cholera has spread through Latin American cities since the 1980s;
- Malaria and other mosquito-transmitted diseases, notably: dengue, hepatitis-A, Nile fever, yellow fever and various forms of encephalitis;
- Tuberculosis (TB), human immunodeficiency virus/acquired immune deficiency syndrome (HIV/AIDS);
- Bubonic plague, which still appears: an outbreak occurred in Surat (India) in the 1990s;
- Debilitation by parasites;
- Respiratory problems due to damp, crowding which favours disease transmission, and air pollution from fuel wood and charcoal;
- Unusually high rates of accidents – burns, scaldings and falls – which result from poor housing, alcoholism, unsurfaced and unlit tracks.

Squatter settlements commonly make use of steep or swampy land rejected by others. In sloping areas landslides beneath or onto housing are a risk whenever rainfalls are intense – at the time of writing (17 January 2003), *The Times* (UK) reported at least 14 dead and seven missing in one day through mudslides which had hit *favelas* (slums) in southeastern Brazil. Steep land is also prone to gullying, and typically has dangerous pathways. Slums cover the hills around Rio de Janeiro (Brazil), Caracas (Venezuela), Bogota (Columbia), and many other cities. Squatter settlements on floodland – like the swamps around Guyaquil (Ecuador), Recife (Brazil), or Lagos and Port Harcourt (Nigeria) and parts of Calcutta (India) – are regularly inundated with sewage-contaminated water. Neighbourhoods where there is inadequate sewage disposal and refuse collection expose inhabitants to disease risk through inhalation of dust and contact with mud, water, contaminated food and insect and rodent pests. The children of the poor seldom have safe play areas, and a fair number are abandoned or obliged to live on the streets, and so run a high risk of contracting diseases or being injured.

Urban environmental problems can be subdivided into those associated with the home, the workplace and the neighbourhood. For poor people the home is often also the workplace, which means exposure of the whole family to noise, fire risk and harmful materials. The employment of child labour is common and children are especially vulnerable to industrial accidents and pollutants. Scattered small workshops and householders undertaking contract work are difficult to supervise and protect and accidents can affect surrounding housing. Innovations in larger factories are more likely to be monitored than those adopted by scattered small workshops, and health and safety education is difficult to deliver. Very few, if any, workers in those workshops have any medical or injury insurance. The problems of poor areas may not be contained there; disease outbreaks, crime and unrest can spill out to affect richer areas, including overseas to developed countries, so it is also in the interest of richer people to resolve threats. Dangerous diseases, like multi-drug resistant TB, dengue and Nile fever have been spreading well beyond slum areas in developing countries to developed countries. In the case of dengue, the carrier – mosquitoes *Aedes albopictus* and *A. aegypti* – recently travelled to the USA and Europe, probably in cargoes of wet car tyres shipped for re-treading.

Urban sprawl

Buildings and other infrastructure expansions are destroying soil resources worldwide. The loss significantly counteracts any gains in food production made through agricultural modernisation; often it is the best and most accessible agricultural land which is built upon. Human disturbance and pollution associated with urban growth impacts on an even greater area. Urban growth removes soil and vegetation 'sinks' for carbon dioxide, and so adds to the problem of possible global environmental change. In developed countries suburban gardens and lawns may help compensate for lost carbon sinks, but in poorer and more arid countries this is less likely unless urban and peri-urban horticulture is encouraged. Urban areas are not adequately mapped at present, making it difficult to estimate the extent of the direct impacts of building and problems like troposphere ozone, smog and acid deposition (Anathaswamy, 2002). Where urban growth is combined with poverty, dependence on fuel wood, dung or charcoal can lead to serious vegetation damage and soil degradation for hundreds of kilometres around, where gangs of collectors strip the countryside of fuel. Where dung is collected soils are denied the return of organic matter and are likely to degrade.

Urban problem areas

Within each urban environment there are particular problem areas – generally the districts where the poor have settled, and the areas affected by industrial activities. Other problem areas are those with soils which make buildings vulnerable to subsidence or earthquake, localities at risk of flooding or landslides, and where polluted air accumulates. Some rural environments are also greatly modified from their natural state and may be heavily polluted, but in urban areas this is a more frequent occurrence. In urban environments poor newcomers are less likely to have established effective coping strategies than poor rural folk and are often isolated from families and communities so have nobody to fall back upon in times of hardship.

At the start of the twenty-first century, "... poverty is suffered by a minority of urban dwellers in richer nations and by the majority in poorer nations ..." (Hardoy et al., 1992: 195). The people who become the poor of cities often have countryside skills which are of little use. Those suffering hardships in urban areas may be more vulnerable than rural folk; theirs is a cash economy, and without enough money survival is a struggle. The rural poor often have famine crops and other fallback strategies which are unavailable to city-folk. This may be partially compensated for by the relative ease of access aid agencies have to cities, and because governments may help urban people because they are the most likely to give them their votes and pose a closer threat if neglected.

Pollution and waste associated with urban growth

There are now many 'mega-cities', with multimillion populations, in countries with limited funds for urban management. Some of these once had adequate urban services but these have been overwhelmed. Consequently, these cities are amongst the most altered, unhealthy and contaminated of Earth's environments. Urban impacts are much wider than is at first glance apparent: sewage and air pollution affect surrounding regions, demand for city domestic water competes with irrigation and wildlife conservation, and energy supply prompts the construction of hydroelectric dams or thermal power stations which further damage the countryside.

Cities are often sited on level plains and valley bottoms so urban expansion destroys some of the best agricultural farmland. A site selected for a cool climate and other benefits centuries ago can have become a trap for smog and a problem to supply with water – like Mexico City, which lies at over 2,500 metres in a bowl-shaped depression. Wherever possible, waste and pollution disposal should be integrated with employment generation, food and commodity production and appropriate industrial development. There are examples of huge expenditures on waste disposal based on developed country experience that are costly to maintain and often a project focus that ultimately shifts problems to a new, end-of-pipe location. For example, the Cairo Wastewater Scheme was funded by the World Bank and the UK Department for International Development (DFID) in the early 1980s to provide improved sewerage that would not become choked with refuse. Unfortunately, while the system worked well it discharged into the Nile with little or no effluent treatment largely because the problem lay beyond the sewerage scheme remit.

A field which seems likely to develop in coming years is that of urban agriculture or horticulture: the growing of crops to employ and feed the poor, which if possible utilises refuse and wastewater. Since the 1980s a Mexican alternative technology development group have been researching the *chinampas* farming system used by the Aztecs before the sixteenth century. This consists of narrow raised fields constantly mulched with mud from narrow canals surrounding them. The canals received the sewage of over 300,000 people and the *chinampas* system is believed to have effectively treated the sewage and enabled it to be used for sustainable and very highly productive agriculture. One expanse of these fields and canals once covered a huge area and fed the city of Tenochtitlán, the present site of Mexico City. Fortunately, an area still survives and functions, which has enabled researchers to assess how the strategy works; one discovery has been the identification of a particularly effective bacterium which promises to be valuable in sewage treatment systems appropriate for developing countries (Ayres, 2004).

Most developing countries have serious traffic pollution; in a few nations, where vehicle ownership is limited, or where technological 'leap-frogging' has led to the construction of improved public transport, there may be fewer difficulties. However, on the whole in developing countries, vehicle numbers are growing fast and the standard of maintenance is poor, so city authorities struggle to find effective pollution controls. Developing country fuel is less likely to be lead-free and catalysers are too expensive to maintain once cars age, so levels of lead, volatile organic compounds (VOCs), troposphere ozone and other pollutants present serious and growing risks. Yet traffic pollution is welcomed by citizens in some cities as a 'sign of progress', and there is little prospect of rapid control unless motorists can be made to pay or funds can be found for effective public transport. Even where there are impressive metro or modern light railway systems, like Mexico City (see Figire 7.1), Singapore, Bangkok, Shanghai and Kuala Lumpur, there are still difficulties dealing with masses of private motorists, taxis and buses. Taxi and bus companies often have to manage on shoe-string budgets because customers cannot pay higher fares.

Sustainable urban development

This has become a much-voiced goal in the last 10 years or so, but it remains a field where rhetoric exceeds actual success (Button and Pearce, 1989; Stren and White, 1992; White and Whitney, 1992). One problem is how to appraise the sustainability of a city, policy or programme (for a review and coverage of scenario accounting approaches to this, see Ravetz, 2000b). Once urban sustainability can be measured attempts can be made to achieve it.

Figure 7.1 *Modern metro system, Mexico City. Low-cost tickets, high-capacity carriages, and frequent services help to relieve congestion above ground.*

Source: Author, 2001.

The pursuit of sustainable urban development may be broken down into:

- Seeking urban policies and plans that support sustainable development;
- Striving to reduce the impact of a city on the environment;
- Sustaining the function of the city.

The 1992 Rio Earth Summit encouraged 6,400 local authorities in 113 countries to produce Local Environmental Action Plans (*Local Agenda 21*s). These were intended to integrate environmental objectives into development plans, emphasising participation, accountability and sustainability, but progress since Rio has been limited. Some cities in developing countries have implemented something slightly different: City Development Strategies which seek broadly what the Environmental Action Plans proposed, but are more pro-poor. Hopefully, these will stimulate a shift to genuine action (Singh, 2001).

Ways of addressing individual urban problems are already available – these will be components in the sustainable urban development strategies that planners and environmental managers are developing. It will be important to select the right components for given situations and to find ways of implementing strategies, paying for them and ensuring they continue to function. Sustainable strategies must also be adaptable. There does seem to be a growing awareness of the need to plan for cities to adapt to global environmental change. Key elements in sustainable urban development are to provide and maintain adequate inputs and deal with outputs so that the city and extra-urban surroundings are not damaged. A clean and sustained city is not sustainable development if the rest of a country or wider environment is bearing the impacts.

Improved urban water supplies

Contaminated water is a major cause of ill health in developing country cities. Overcoming the problem depends on finding adequate supplies, treating and conveying them safely to consumers and carefully disposing of waste water. Poor people may already rely on private sector suppliers, and adequate piped water supply systems and sewerage improvements for some may come via that route, rather than city authorities or foreign aid (Cairncross and Feacham, 1983: 89). Another route is to support community self-help schemes.

Improved urban waste disposal

Degraded areas of cities which are ripe for redevelopment have come to be termed 'brownfield' areas in the developed West, and are often former industrial sites. Redeveloping such land may mean very expensive decontamination work. In developing countries that is less common, and the problem tends to be substandard dwelling sites to clear or upgrade. Aid agencies often finance construction but offer little recurrent funding. Consequently, a problem for urban environmental management is fundraising for ongoing costs, repair and modifications, because in poor areas the locals are unlikely to be able to pay much tax.

Community-based approaches are currently seen to have promise for urban improvement, which means environmental managers must have at least a passing familiarity with social issues and with participatory approaches (Hasan and Khan, 1999). Slum upgrading in developing countries is now often conducted by giving local people the materials for housing and road improvements, and providing pipes and skilled labour for sewerage. In many urban areas of developing countries, excreta and refuse are left around houses, often on vacant ground or street sides – a most unsatisfactory situation. If there is any organised removal it is by cart or tractor and trailer; often there is not, and removal may not mean safe disposal. Some settlements do convey refuse and excreta ('nightsoil') to surrounding agricultural land, composting or biogas production facilities; when well-managed these can be effective, affordable and relatively safe (Mara and Cairncross, 1990).

Unfortunately waste and excreta may be so contaminated with heavy metals and troublesome chemicals that opportunities are limited for recovering solids to use as compost or using the effluent for irrigation. Poor districts of developing country cities are also likely to discharge sewage contaminated by small-scale industrial activity – especially oil and solvents.

Where there are sewerage facilities in developing countries these are generally Western-style water-based systems offering only rudimentary treatment. Increasingly, cities have been extending low-cost versions to poor areas. A popular approach has been for governments and non-governmental organisations (NGOs) to support the installation of cheap pour-flush (bucket-flush) sewerage systems. By supplying moulds for making the toilet slabs and subsidising piping, costs can be kept down. However, slums spring up unconstrained by any planning, there are no regular roadways and housing is scattered up hill and down dale – consequently, it can be very difficult and costly to provide services like piped water, mains electricity and sewers. Provision of toilets and sewerage (pipes) is less of a problem than the safe disposal of sewage (effluent); many systems simply discharge into streams or the sea or are connected to septic tanks. While this can be satisfactory for smaller settlements and in certain geological conditions, and does give less opportunity for flies and vermin to breed and come into contact with people, it frequently results in severe groundwater and surface water contamination. Also, water-based sewerage systems are easily overwhelmed during heavy rainstorms, especially where there is poor provision for surface water drainage. If possible, sewage-contaminated water should be kept separate from 'grey' water – washing water and storm runoff – because the latter is easier to process and can be allowed to escape when there is heavy rain. Where there are many settlements along a river and if sewage reaches shallow and confined seas or lakes, pollution is likely to be bad.

Authorities can help healthcare by ensuring standing water is regularly treated to prevent fly and mosquito breeding; for example, polystyrene beads can be floated on water surfaces, oil and pesticide can be sprayed, or suitable fish can be introduced to eat the insect larvae. Where dengue is a problem inspections and fines for those caught with water-filled litter on their property – the potential breeding sites for the mosquitoes – should help reduce transmission (see Figure 7.2).

There are alternatives to water-based sewerage. The problem is making effective, affordable systems that people are willing to use. People are familiar with flush

Figure 7.2 *Team spraying insecticide on houses (Northeast Brazil). The state department regularly sprays pesticide to help control malaria, yellow fever, dengue and Chagas disease. Close proximity to the river means that aquatic life will probably get contaminated.*

Source: Author, 1982.

toilet, water-based systems – perhaps not at first-hand, but they see them on TV – and more companies are currently geared up to manufacture, sell and install them. Waterless systems have to overcome some consumer resistance.

Where settlements have no effective waste or sewage disposal it may be possible to establish community-organised composting, using simple carts to collect nightsoil and refuse. Often the poor have already established urban agriculture, fuel-wood production or aquaculture, and may use sewage and food waste for irrigation or as a way to feed poultry, fish and pigs. In the early 1990s Shanghai, in China, was supplied with over 80 per cent of its vegetables from urban and peri-urban farmers using nightsoil (Hardoy et al., 1992: 139). Around Calcutta sewage is added to fish ponds to help feed the fish and to grow water hyacinth as fodder for other livestock; similar activities take place close to many other developing country cities. It is better for authorities to assist and monitor such activities than pretend they do not happen. Better to pay to chlorinate sewage outflows, than have it applied untreated to vegetables sold in city markets. A flourishing urban and peri-urban agriculture sector should improve food supplies, help dispose of wastes and generate livelihoods for people otherwise unlikely to find useful employment (Deng, 1992). A sustainable strategy linking waste and agriculture must be monitored by people with a broad knowledge of pollution, healthcare and so forth. Activities like chicken or pig rearing near cities can pose serious threats even if waste recycling is not involved – avian flu and severe acute respiratory syndrome (SARS) seem to originate from contact between fowl, livestock, insects and humans in such conditions. Urban environmental managers must keep a close eye on urban and peri-urban agricultural activities.

The urban poor generate much less refuse that rich city folk and commonly build with waste materials, so that impacts from cement manufacture and brick making only appear when authorities upgrade slums with such materials. With limited consumption of consumer items there is far less packaging waste per capita than would be the case in richer countries or more affluent areas of their own city. Much of the waste is collected and recycled in some way to generate income or serve the households.

A number of developing country cities have begun to aid the poor to sort and recycle garbage (Perera and Amin, 1996). For example, concrete sorting platforms can greatly help; most important is to try and improve basic health and safety facilities by providing washing facilities, supplies of boots, gloves, tetanus injections and health treatment. One of the most basic improvements is to organise regular refuse collection and removal to a site away from habitation. This does not need sophisticated garbage trucks: tractors and trailers will often suffice (UNCHS, 1988). In Cairo, the authorities supported traditional waste collection with improved donkey-carts for areas with narrow streets and it proved highly effective.

Housing and neighbourhood improvements

Squatter housing (slums, shanties) are little proof against vermin, damp, cold and severe storms, and are prone to fire. Lightly built slums are unlikely to kill many by collapse during earthquakes, but they suffer from fire if the quake happens when people are cooking; however, where people use heavier mud construction, collapse may pose a danger to inhabitants. Where funds are available, governments can make considerable improvements to people's well being by supplying materials for housing improvement: building blocks and roofing sheets can cut fire risk and damp; cheap doors and windows can help cut malaria transmission and crime.

Urban transport improvements

Developing cities rely a great deal on bus and taxi transport and, partly thanks to leapfrogging, some have excellent systems. Many cities, however, lack adequate public transport and tend to have few pollution control measures, so traffic pollution is a problem. In South Asia and parts of Southeast Asia three-wheeled autotaxis are common; whilst they offer cheap and effective transport, use little fuel and cause less congestion than full-size taxis, they are noisy and seldom equipped with pollution control devices. Some cities have adopted rapid transit systems: state-of-the-art trams, metro or rail systems. Mexico City has an impressive metro system, and Kuala Lumpur and Singapore have modern elevated tramways. Without effective transport there is the risk workers will crowd around manufacturing areas, which can mean safety risks, and inner city districts may develop tenements.

Charging and quota access initiatives have had limited success in excluding private motor cars.

Improving energy supplies

Access to electricity for any but the rich is patchy in the developing world. Some of the poor illegally connect to supply networks; many make do without, using fuel wood, charcoal, kerosene or bottled gas. Accidents are common and the fuels cost a fair proportion of people's incomes. With urban smog and often overcast conditions, cooking with solar power using parabolic reflectors is unlikely to be practical. Household photovoltaic (PV) supplies, which are small solar arrays providing limited amounts of electricity to households, schools or community services, do look likely to spread. PV systems have been falling in cost, and are now being adopted by the 'more affluent' poor as a means of powering a TV and possibly some limited lighting, which allows extended hours of working and evening study for children, but will not power cooking stoves. PV is used by local cafes or bars to offer TV and telecommmunications. Local authorities and private companies are also starting to install PV public telephones. Cheaper, robust systems could gain even wider use.

As PV and TV spread there are opportunities to reach more people via the media; a priority should be to educate and mobilise local action to improve environmental management (Abbott, 1996; Toteng, 2001). It is not only important to educate developing country citizens about the need for better environmental management; their governments, administrators and planners must also acquire improved awareness. The poor frequently shift to kerosene, because bottled gas is less readily available and more costly. Biogas is difficult to organise for small urban households, but might be feasible at a neighbourhood scale if linked to excreta and refuse treatment. Unfortunately, not all settlements will have suitable waste. So far such biogas adoption has been mainly in China and India.

For decades a growing problem in many rural and urban areas has been the supply and use of fuel wood or charcoal (Leach and Mearns, 1989). Merchants generally take over once a settlement grows much beyond small township size, and they can strip huge areas of vegetation. Hardoy et al. (1992: 110) reported that Delhi shipped in 612 tons of fuel wood a day by rail alone in the early 1980s – and it is a city with abundant cheap coal, and reasonable access to kerosene and liquid propane gas. Settlements in Asia and Africa cannot continue to deforest vast areas for fuel wood and charcoal; alternatives must be developed. Sustainable fuel wood plantations may help ease this degradation but it remains to be seen whether in their present form these are truly sustainable and beneficial, and whether the planting can keep pace with demand. Plantations are often eucalyptus stands on what was once common land which means people get relocated and the trees generate a range of unwanted environmental and economic impacts. Sustainable plantations of fast-growing plants – willows, reeds or aquatic plants – possibly fertilised with sewage can be converted to biogas, charcoal briquettes or woodchips to burn in district generation stations. The ideal is to dovetail sewage and refuse disposal with energy supply or composting.

Those in richer districts of developing country cities and towns may well increase energy consumption and per capita refuse in coming years. Attempts to tax energy or other basic commodities hits the poor more because buying them takes a greater portion of their income. The rich spend a lower proportion of their income on such commodities and may be more able to evade tax.

Who pays to care for the urban environment?

Much of the investment to pay for improvement of cities in developing countries will have to come from outside those countries (Leitmann, 1996). More and more cities in developing countries are servicing richer countries: for example, Indian cities like Mumbai (Bombay) and Bangalore process tax documents, provide telephone directory enquiries, air ticket sales and accounting, maintenance of bank accounts, credit card helplines, service insurance claims and so forth; Shanghai and other Chinese, Korean, Taiwanese and other Third World cities manufacture more and more of the consumer items purchased in rich nations. Such activities could help pay for improvements if it can be effectively taxed. There could be a downside if losses of clerical jobs in developed countries generates ill feeling and reduces aid donations. It is sometimes possible for tourism and horticulture to generate funds – a number of African, South East Asian and Latin American urban areas produce cut flowers and vegetables for airfreighting to the supermarkets of developed countries.

Cities in developing countries and global environmental change

Environmental change, both natural and human-induced, easily affects the food and water supplies to cities (McCulloch, 1990). However, in the next few decades the greatest impacts on urban areas may well be through rising sea levels. Many cities are sited on coasts and are vulnerable: for example, Alexandria (Egypt) and Dacca (Bangladesh). In addition to flooding and storm damage, rising sea levels are likely to cause penetration of saltwater into estuaries and the displacement of fresh groundwater. Already some countries have started installing coastal protection bunds, although these are no real solution and probably do not justify the huge cost. Bangladesh flood defences have caused considerable debate. The alternative to Bangladesh and aid agencies spending huge sums on flood protection embankments which may be ineffectual or even dangerous is for planners to discourage vulnerable land use in areas likely to flood and to provide strong flood and storm refuges and early-warning systems as a civil defence measure.

Global climate change may well alter rainfall, resulting in changes to urban water supply; the result might be shortages or more flood risk, and altered susceptibility to landslides if there is more intense rainfall. It is not unusual to find that countries have building regulations based on those of colonial powers with non-tropical environments, so they are out of step with needs. Also, storm drainage and water supply systems are likely to be built to tight budgets; these factors mean there is little flexibility to cope with environmental change. Environmental change may well have urban health impacts, but warnings are difficult to assess because many complex factors are often involved. Predictions so far are too often based on guesswork: for example, global warming is widely assumed to be likely to cause a spread and increased incidence of malaria. Yet in Europe before the 1870s malaria was common, although conditions were cooler for much of the last 350 years than now. Factors other than climate seem to explain the reduced transmission in Europe.

One of the more likely impacts of global warming could be shifts in energy use and consumption of goods as rich and poor countries implement economic controls and introduce new technology to honour agreements over carbon emissions and adapt to climatic change. Environmental management must forecast and prepare for changes, and it must try to do so as objectively as possible.

Industrial pollution

Some pollutants are predominantly released in urban environments, some are associated more with rural activities, others contaminate both. Making a clear separation between them is not easy, but as far as possible agricultural pollution is discussed in Chapter 4; there is some coverage of water pollution in Chapter 3, and atmospheric pollution in Chapter 6.

There is no universally accepted definition of pollution; the following is reasonably satisfactory: the introduction by humans, deliberately or inadvertently, of substances or energy (heat, radiation, noise), into the environment resulting in a deleterious effect (O'Riordan, 1995). Contamination is the presence of elevated concentrations of substances in the environment – food, water or whatever – which may be a nuisance or harmful. Pollution involves contamination, but contamination does not necessarily constitute pollution if it is inoffensive and harmless. Nature can generate contamination and pollution – for example, radon gas emissions, volcanic ash and gases, toxic algae blooms and many other things. Basically, a pollutant is something in the environment that causes harm or offence. New pollutants are constantly being created and as research progresses, materials now believed safe may be found to be a threat. A European Union (EU) study in 2004 found 86 per cent of 2,500 chemicals marketed in large volumes had inadequate data available on their toxicity.

Pollutants vary greatly in their ease of dispersal; some spread little and others globally, some may move from air to water and back to air, or from water to food, or in many other ways. While some pollutants easily degrade after dispersal to harmless compounds, others turn into harmful materials or resist breakdown and persist for a long time. Genetically modified organisms (GMOs) may continue to spread and evolve long after an escape has been stopped. In addition to toxic and hazardous pollution, there is also that which is in some way a nuisance, has aesthetic or economic impact – litter may not be particularly damaging to health but it can depress people, discourage tourism and drive down property prices.

Determining 'safe' levels of contamination can be difficult because some compounds or energy emissions affect organisms at very low exposure and have insidious and indirect, often cumulative, impacts. Certain highly toxic or carcinogenic materials, e.g. pesticides such as dichlorodiphenyltrichloroethane (DDT) or polychlorinated biphenyls (PCBs), may be concentrated by physical processes like stream currents or interception of mists on vegetation to form 'hot spots', or by biochemical processes, notably as organisms feed and are fed upon by predators. The latter process of bioaccumulation (biomagnification) can mean that a low background level of a pollutant in the environment becomes magnified enough to harm higher organisms. It is also possible for pollutants to be accumulated over time in soil, often bonded to clay minerals, or in winter snow cover. In the former case environmental change may trigger sudden release, and spring thaw in the latter causes a surge of pollution, situations often described as time-bomb effects.

The mode of release varies a lot: some pollutants disperse from point sources like a single chimney, some from large areas, e.g. smoke from a forest fire, some from linear sources like highways. The emission may be a single brief event like an explosion, periodic, sporadic or continuous. Measuring pollution and setting standards to aid monitoring and to try to avoid pollution is part of the job of pollution control specialists.

Waste can be defined as movable material that is perceived, perhaps wrongly, to be of no further value; once discarded it is a nuisance or a hazard, but may turn out to be a valuable resource. As with pollution there is a huge diversity of types of waste, some of which can break down with time or react with the environment or other waste to release pollutants. For management purposes it is possible to subdivide waste into toxic and non-toxic, or according to origin: industrial, agricultural and domestic (refuse, garbage or trash). Alternatively, it can be split into organic and inorganic. For treatment or recycling, waste generally needs some form of sorting, which can be difficult and costly.

Pollution and waste management can focus on:

1 Prevention: fully proactive avoidance;
2 Treatment pre-release;
3 Mitigation post-release: more likely to be countermeasures, dispersal and dilution or treatment of symptoms, rather than cure of the pollution – 'end-of-pipe' solutions;
4 Amendment or rehabilitation post-release: chemical, physical or biological treatment intended to collect and dispose of the pollutant or waste or to alter it to something less harmful;
5 Recovery and recycling, pre-release or post-release: the material is converted to something useful – raw material for manufacture, heat, compost for agriculture, feed for livestock and so on. This may partly or wholly pay for itself, perhaps even provide a profit. Ideally environmental managers seek to integrate waste generation and other activities to be mutually beneficial and cut pollution (Anandalingam and Westfall, 1987).

Pollution and waste may present an immediate problem or become so only after considerable delay. Those who generate pollution and waste may not suffer the effects, while others, possibly well offsite or removed in time, do. There are situations where industrialists made profits some time in the past and present-day citizens have to pay to clean up because the polluter is untraceable or long dead. Problems also arise when pollution affects common resources and crosses borders.

Various groups are affected differently by pollution and waste, especially in a multicultural society, where dissimilar lifestyles means varied exposure and different concerns. For example, studies in the USA have shown that the poor accumulate much higher levels of pesticides in their bodies, because their diet is more likely to be contaminated and they are exposed during menial work and through poor housing (Hurley, 1995). In the UK it is quite noticeable that poor people tend to live in the east of towns and cities, downwind of most pollution sources. Children and women may be differently exposed and physically more vulnerable than the male working population for a variety of reasons. So it may be expedient for environmental managers to concentrate on different groups: bystanders and the public, through public health measures; workers, through safety at work; consumers, through consumer law and public health measures; and various age, gender or ethnic groupings. It may also make sense to adopt a sectoral approach, concentrating on urban, industrial, water-related, agriculture-related, tourism-related and warfare-related pollution and waste as well as biotechnology, genetic problems and radiation. Pollution and waste are a problem in any political situation: capitalist or communist, centralised or decentralised, Western or non-Western.

Sustainable development demands effective management of outputs as well as inputs, and action now, rather than leaving problems for the next generation, which has often been the habit in the past. The outputs of development are pollution and waste. In an ideal world where people were seeking sustainable development, pollution and waste management would not be neglected; however, in all countries, especially poor developing nations, decision makers seek trade-offs between economic growth, food production and employment generation on the one hand, and expenditure of scarce resources on environmental management on the other. One such situation is when a developing country has to decide between using an affordable but environmentally damaging pesticide or relying on more costly, and possibly less effective, 'safer' solutions to fight a serious disease like malaria. The longer-term and global view would be to urge use of the safer, expensive and less effective control, but with limited funds and an urgent healthcare problem the poor nation can see the benefits of ignoring environmental advice. The rational solution would be to target aid to pick up the extra costs of safe malaria control. Environmental managers can warn aid donors of such situations.

Exports from some developing countries may contain contaminants in sufficient quantity to pose a threat to consumers; sales are then likely to be controlled. Because poorer countries generally have less stringent environmental legislation and abundant cheap labour, they may host relatively dirty technology which exposes many workers and other citizens to risk. Many developing countries are tropical or subtropical, which means polluted water bodies can be rapidly depleted of oxygen through high chemical oxygen demand (COD) and biochemical oxygen demand (BOD) – see Glossary.

Broadly, developing countries are likely to give priority to health improvement, employment generation and foreign exchange earnings; developed nations are more likely to have funds that also address biodiversity conservation, aesthetic quality and nuisance issues. Industrial projects or large tourism developments can often be forced to install pollution and waste control measures. A common developing country problem is the large and rapidly growing city where the civic authorities have scarce funds for upgrading sewerage, waste collection and processing, or for adequate traffic controls. While it is difficult to accurately generalise, Box 7.2 lists some current trends in developing country pollution and waste management.

Box 7.2

Current trends in pollution and waste management in developing countries

- Increasing adoption of the polluter-pays principle. A move toward more open and proactive planning and management; associated with this is more media oversight, 'whistle-blowing' and the use of tools like EIA.
- The spread of guidelines, regulations and legislation which requires or encourages the release of information on risk and status of pollution and waste.
- The introduction of more and better environmental quality and human well-being monitoring standards.
- Improving tools for monitoring, so that these are cheaper, more accessible and more accurate: geographical information systems (GIS), remote automatic sensors, data transmission from sensors to processing centres, etc.
- Improving legislation to define, monitor and control pollution and waste and some progress with legislation designed to discourage and require the polluter to pay; there is still a long way to go before things are anywhere near satisfactory.
- More voluntary control: the adoption of environmental standards accreditation, self-regulation and joint agreements between potential polluters and agencies to aid regulation.
- The spread of an ethos which seeks to address pollutants before discharge into the environment: 'clean-up-later' is becoming less acceptable but is by no means unusual in developing countries.
- The spread of the Internet: it is less easy to conceal pollution and waste nowadays.
- The end of the Cold War has improved the exchange of information, and possibly released some funding for environmental management.
- Rapid advances in biotechnology, the constant synthesis of new chemicals and pharmaceuticals, and the spread of radioactive sources present new challenges; ongoing vigilance is needed. Upgrading of information is vital. With new developments impacts may be unexpected and there is little hindsight knowledge.
- Since the late 1980s some countries have imposed austerity measures, often due to structural adjustment programmes; this may have cut funds to pollution control and waste management and may force industry to try to save on treatment costs.
- Some countries are rapidly developing their manufacturing sectors and are burning coal and other polluting fuels and have increasing numbers of road vehicles with poor pollution control measures: South Asia, Southeast Asia and China in particular have serious and growing smog and acid deposition problems which also cross national boundaries and have global impact.
- Global controls on pollution – measures to combat stratospheric ozone damage, global warming and the release of toxic chemicals like pesticides and PCBs to the global environment – are unlikely to be adopted by developing countries unless there is environmental management aid from richer nations.

Source: Author

In most countries there is a tendency for rich companies and individuals to benefit from development, while poorer locals suffer impacts like pollution and waste. Poor people are also less likely to turn to litigation, or win reasonable compensation or protection if they do (see Box 7.3).

Box 7.3

The Bhopal tragedy

In 1984 an accident at Union Carbide's Indian Subsidiary's pesticide plant in Bhopal gassed many people. It is widely accepted that this was the world's worst industrial accident. Study of this tragedy reveals why it happened and how to reduce the chances of similar accidents. There is also an ethical aspect relating to safety measures and compensation for accidents, including the question of whether a human life is worth the same worldwide.

The accident happened at night, when a faulty valve let water into a tank holding 40 tonnes of reactive methyl isocyanate (MIC) used for insecticide production. A runaway reaction occurred, gas got past safety equipment and escaped as a toxic cloud (one compound may well have been cyanide). Close to the plant lay two settlements of low-cost housing. In developed countries safety measures are much tighter.

So, was the disaster an unforeseeable event? Difficult to prepare for? Or was it caused by poor design, mismanagement, faulty planning, weak legislation, negligence or sabotage? Was the cause rooted in local administration and management, national causes, the activities of an American parent company or a combination of factors? The victims were mainly bystanders, rather than workers or consumers. The shield for employees is mainly through safety-at-work measures; for bystanders it should be local and international planning, management and legislation; consumers can be protected through product controls. Without enforced standards commerce will be tempted to cut corners to be competitive.

Multiple safety systems failed; apparently a key component, which should have cooled the MIC tank enough to reduce the risk of runaway reaction, had been drained of refrigerant. Some reports suggest this was for repairs to ensure a safety inspection would be passed, others claim it was to save relatively small sums of money. Another claim is that maintenance was a problem in humid tropical conditions, where skilled engineers were scarce and delivery of spares delayed. Bhopal had no empty tank to catch a runaway reaction. It had one flare tower – designed to burn off toxic escapes – but this was shut for repair. A scrubber – a unit where gas would be sprayed with caustic soda solution – was overwhelmed. The gases and boiling liquids generated were roughly 100 times more than the scrubber could handle, which suggests design flaws.

The Bhopal plant opened in 1980. Clearly, any authority reviewing the proposed plant must have weighed risks against the need for jobs, the benefits to agriculture,and the possibility that the development would attract other investors. In 2002 estimates placed deaths at about 8,000, with at least 150,000 seriously injured (figures are rough because many victims were unrecorded and delayed effects constantly appear). Litigation has been complex and continues. The Indian Government initially sought damages of US $3.3 billion. In 1989 the Union Carbide paid US $470 million to a trust fund for victims, but claimants appear to have received only about US $500 each.* One aspect of the case was the arrival of 'ambulance-chasing' lawyers, offering no-win-no-fee deals but taking a portion of any successful claims. At times the Indian government has intervened to try to aid litigants and co-ordinate claims. Out-of-court offers by Union Carbide of US $200 million, and later US $350 million were rejected by the government of India in 1985. A class or group action was initiated by some Bhopal survivors in New York State in 1999, which led to the release of company documents in 2002 that has renewed debate. The USA parent company retained enough control (a 51 per cent holding) to be able to influence activities, but claimed to have exercised a 'hands-off' relationship with its Indian subsidiary, which it maintains was responsible for design and running. In 1972 Union Carbide seems to have been concerned it would need to issue so many shares to raise funding that it would reduce its control below a sufficiently dominant level, and this some suggest may have reduced investments, which cut provision of safety measures.

What could have been done to reduce the risks, apart from better equipment and plant monitoring? Bhopal might have enforced strict land-use zoning to keep settlement at a safe distance, but rapid urban growth and a need for poorly paid workers to live close to work eroded such hopes. Cheap public transport might have helped by making it possible for workers to live well away. Early-warning systems could have woken those nearby and many could have evacuated; some claim simply lying down with a wet cloth over the face would give reasonable protection, but the public was unaware. Had hospitals known that an MIC escape was likely they could have trained in appropriate first aid and stockpiled suitable drugs and adequate oxygen, at the expense of Union Carbide. After the escape there was confusion over the cause of the respiratory distress and the best treatment. There have been claims that the authorities failed to warn affected pregnant women that there was a risk of birth defects and that contraception must be used until people were sufficiently de-toxified, which takes time.

Bhopal has encouraged more caution from developers, better monitoring from governments and international agencies and more awareness among the media and NGOs. Since 1977 the United Nations has worked on a Code of Conduct on Transnational Corporations which includes clauses covering environmental and safety responsibilities.

* The compensation figures should be compared with North American or European payments; for example, an ammonia leak from a railroad accident on the Louisville–Nashville railway line in the USA which killed two people resulted in the payment of US $52 million. Compensation to those simply briefly evacuated and stressed during the Three Mile Island nuclear accident, also in the USA, cost Metropolitan Edison US $33 million.

Source: Consumers Association of Penang, 1984; Ayres, 1987; Weir, 1988; Bennett, 1992: 174-197; Shrivastava, 1992; Jasanoff, 1994; Morehouse, 1994; Lapierre and Moro, 2002; Mackenzie, 2002

Quite a lot of manufacturing and commodity processing in developing countries is conducted by many small family businesses, typically scattered through residential districts or in villages. These have little to spend on pollution and waste management and are difficult to monitor and control. The benefits from effective waste treatment in terms of improved labour productivity, less demand for healthcare and more pleasant environment for attracting tourism revenue may prompt a federal government to contribute to environmental management.

Some pollution and waste treatment strategies and technology do not suit developing countries or the tropics. One could argue that water-borne sewage disposal is inappropriate if there is growing water shortage, and because sewerage is costly. There have been some innovations to make water-borne disposal more appropriate: such as simplified sewerage that resists blockage. However, low-cost and effective techniques need further development: for example, sewage farms – orchards or non-food crop farming using sewage for irrigation – are seen in developed countries to be obsolete, but could be improved and promoted in developing countries.

Few water-borne sewage systems in developing countries have reasonable treatment; most simply discharge into the nearest river, sea or marshland. The consequences can be dire; however, treatment to all but the most basic standards is presently costly and prone to breakdowns. Alternatives are being developed, one being biological waste treatment: reedbeds or channels with water hyacinths (*Eichhornia crassipes*) or algae, which remove pollutants and can then be harvested for fuel, construction material or compost. Another approach is to have a bucket system and daily nightsoil collection for composting. More acceptable in the long run would be to install toilets that compost on site and only require emptying of relatively inoffensive debris once a year or so.

Waste recycling has been improving in some developed countries and tends to rely on household sorting or the application of quite costly technology at collection points. Developing nations currently dispose of most waste to 'dumps', where the informal sector sorts much of it for recycling. Simple garbage collection and sorting can be quite efficient and relatively low cost, and poor people may welcome the employment opportunity. Unfortunately, it has sprung up informally, is unhealthy for those involved and seldom provides a reasonable income. A typical waste dump in a developing country is usually all too apparent; often there is no adequate burial, leading to wind-blown debris, unpleasant smell, smoke from spontaneous and set fires and vermin. The approach can be upgraded to achieve better burial of waste, reduce littering and improve sealing of sites to prevent leakage of pollutants, and efforts can be made to assist garbage workers and those picking over the material to reduce health risks. Improved recycling should give some cost recovery. It may also be possible to improve collection vehicles and find ways to access narrow and unmade roads for refuse collection. The authorities in Curitiba (Brazil) provide communal end-of-street refuse containers where vehicle access is difficult and incentives for people to deliver waste to these and for sorting recyclable materials. This includes the issue of tokens redeemable for public transport, entertainment and other services.

Provided the problems of harmful compounds and dangerous organisms can be overcome, domestic refuse and sewage are potentially valuable commodities, especially in areas where soils lack organic matter and chemical fertilisers are expensive and likely to pollute the environment. Composting, if well managed, should generate temperatures high enough to kill most harmful organisms; reducing heavy metal and chemical contaminants is more of a problem. A number of countries are starting to explore ways of using waste for agricultural benefit, through composting and earthworm digestion. Biogas generation may be able to make use of contaminated waste to provide methane for cooking, heating or electricity generation; however, it demands a steady supply of suitable waste and not all settlements can provide this.

Oil palm, sugar and rubber processing, fruit canning, alcohol fuel production and fish or meat packing all generate waste, which can do tremendous damage to rivers and other water bodies if it is not adequately treated. Less obvious but with serious impact in some regions is the illicit production of narcotics; rivers in a number of Latin American and Asian states are suffering because of the acids, detergents and other materials discarded during cocaine and heroin production, plus herbicides if the authorities try to kill the raw materials in the fields.

Urban and rural areas often rely on wells or are supplied through low-pressure pipes that can become contaminated from the soil, leaking drains, chemical and oil escapes. Where sewers choke with sediment it is possible to install new ones, or reline existing systems, with a cross-section that is egg-shaped, narrow at the base, or stepped, to ensure that even when flows are weak the debris are swept along better than with normal-shaped conduits. This reduces the risk of blockages and standing water that might seep into the ground. Inadequate sewage disposal is a key factor determining well being in developing countries and calls have been made for huge sums of aid to improve sanitation and to protect drinking water supplies.

The export of waste and pollution hazards to developing countries

Some export of waste and pollution hazard is unintentional, but there is also deliberate transport of dangerous material to try and take advantage of less strict controls. Often the problem is the result of poor information exchange. Of course, there are situations where profit motives and a cavalier disregard for the well being of people dominates. One way to help avoid problems is to disseminate adequate information about the risks, and environmental managers have worked with various international agencies and NGOs to do this. Access to information assists NGOs and citizen groups to be alert for problems and aids developing country government agencies in monitoring and responding to problems. Awareness about the transfer of hazards to developing countries has improved since the late 1970s, but the problem is still not adequately resolved and monitoring remains poor (Ives, 1985; Clapp, 2002; Strohm, 2002).

The export of hazards takes a number of forms:

- Illicit disposal of hazardous waste in developing countries, which may be unaware of the risk, or where officials judge state or personal profits to be a reasonable trade-off against risk to citizens. The dumping may sometimes be undertaken by unscrupulous private individuals or commercial bodies without officials being aware of it.
- The gift of or sale of unsafe goods to developing countries – material which is likely to be prohibited or restricted by regulations in developed countries.
- The gift or sale of material to developing countries which is likely to be unsafe: for example, powdered baby milk formula may be fine in a country where all have access to clean drinking water, but elsewhere it could be harmful compared with breast-feeding. Fungicide-treated seed is potentially of great benefit, but in a few countries – like Iraq in 1972 – people have consumed it in times of hardship or because of poor warning labels, and consequently have suffered mercury poisoning.
- The establishment of factories or services that present risks that would not be accepted in developed countries. When a multinational company is faced with decommissioning an old plant, it may decide it can still earn a profit and provide a developing country with foreign exchange earnings and employment if it ships the old plant or constructs a new one to standards

that are 'adequate' for a developing country. Some activities undertaken by developed country agencies in developing countries would not be tolerated in their own nations: for example, the spraying of defoliant on illicit drug farms.

- Genuine errors: aid or commerce spreads new technology which subsequently proves to pose risks. Possibly the techniques are safe enough in a developed country, but in developing countries things differ. There may be child labour, and workers and bystanders are more likely to be undernourished, making them more vulnerable to pollution than well-fed Northerners (Bull, 1982). Bangladesh recently tried to sue the British Geological Survey for failing to recognise the threat of arsenic in groundwater when there were surveys of aquifers 30 years ago. These have been tapped by boreholes and people are now being poisoned. The Survey pleaded that the threat was not apparent at the time of surveying, and that the aim was to quickly and cheaply provide drinking and irrigation water that was safer than the contaminated shallow groundwater. The plea was upheld by the courts, but the question remains – is development in the South cautious enough?

Some hazards continue to be exported, yet knowledge exists to reduce or eliminate them; the issue is often one of profit versus human well being, biodiversity conservation or environmental quality. Asbestos is an example: in developed countries the risks are now known and legislation has caught up enough to discourage the threat, but still not give those contaminated in the past sure and adequate compensation. In developing countries asbestos is still used in a way that is often obscenely dangerous. It will take lobbying from citizenry, workers unions and NGOs to achieve better controls and ensure people are properly aware of the danger. Reality is that people know there is risk, but jobs are scarce and the alternative to employment is hunger and illness as real as any 'textbook' hazard.

Many developing countries have little or no trade-union activity, so there is one less potential control or catalyst for the improvement of worker awareness and public safety. Less-scrupulous companies may locate hazardous activities in such situations because there is little chance of restrictive practices and strikes. There are developing countries which promote their lack of unionisation and easy-going environmental and employment safety laws to try and attract foreign factories, so improvement may face marked resistance. A few rich nations have started to send waste to developing countries for sorting, raising the question: is this export of pollution, or an acceptable source of foreign exchange and raw materials if it can be managed well?

Safeguards which suffice in developed countries may not work: where it is hot workers are tempted to shed protective clothing and masks; chemicals may degenerate in storage under tropical temperatures, forming unexpected hazards; warm and moist conditions are likely to damage warning labels and seals; and illiterate workers may not be able to read or may fail to grasp warnings.

Common industrial and urban pollutants

Some pollutants are clearly industrial or urban, others relate to agricultural activities, many are less easy to pin down. Disease transmission and tourism-related pollution are neither exclusively urban, nor wholly rural, and they are not industry related.

Asbestos

Asbestos is still widely used in developing countries for construction and manufacture of products such as roofing sheets and pipes. It is mainly a workplace and localised problem, but those affected can suffer badly. Control is still far from perfect in richer nations and cleaning-up material installed decades ago presents a challenge; in poorer countries there is a much lower level of awareness, and funding for better handling is limited. A number of developing countries mine and process asbestos for export, some with poor health and safety procedures. Bangladesh now generates foreign exchange by dismantling a good deal of the world's old shipping – and asbestos dust is one of the hazards associated with such activities.

Blue and white asbestos pose the greatest threat; brown asbestos is less of a hazard. Those directly in contact inhale or ingest the material and carry it on their clothes to bystanders and families. Chronic debilitation, in the form of fatal respiratory disease and cancer, can manifest decades after exposure, which makes proving the links to the polluter problematic. Most of those contaminated in developing countries stand little chance of getting adequate compensation.

Much can be done to educate the workforce and public of the dangers, so at least there is more care when working with the material. However, in hot and humid conditions safety equipment may not be worn, and where employment is difficult to come by workers are likely to take risks and unions are absent or less likely to sufficiently protect employees. Trade unions (if active), consumer protection bodies, and other NGOs offer channels for educating employees and management about dangers like asbestos, and can pressure for better safety measures.

Polychlorinated biphenyls

PCBs cause damage at very low concentrations and may persist without degrading for a long time. Already there is widespread contamination, which is a transboundary as well as a local or regional problem. Treatment of PCBs before release into the environment is difficult and costly. Monitoring is important, and it is in the interests of all nations to support this. The levels in oceans are already a matter for concern, leading to contamination of some fish stocks, and might account for recent deaths of marine mammals and some coral reefs. The source is mainly industrial activity.

Persistent organic pollutants

Some of the most dangerous compounds are the persistent organic pollutants (PCPs). Some of these cause cancer, and nervous-system, reproductive-system and immune-system damage, even at very low concentrations. Because they are persistent, organisms can accumulate them and they are difficult to remove from the environment. Some are derived from industrial activity, others are pesticides used in agriculture. PCPs include dioxins, PCBs (see earlier), furans, DDT and various organochlorines. The worst 12 of these PCPs are the focus of the 204 Stockholm Convention on PCPs. This Convention legally binds governments to immediately cease production and any release of existing stocks. At the time of writing over 50 countries had signed the Convention.

'Gender benders'

A number of pollutants, including low levels of some pesticides, act as endocrine disrupters (androgen disrupters or 'gender benders'), mimicking animal and human hormones. There is evidence that this affects human and wildlife reproduction, and possibly causes cancers, even at very low concentrations in the environment. The effects on wildlife and possibly humans include lowered sperm counts and damage to developing embryos, resulting in a predominance of one sex and birth defects. So far it seems mainly to affect wildlife in certain rivers, wetlands and lakes. In a number of countries fish and reptile populations have markedly shifted gender characteristics, sometimes enough to threaten survival. In Europe and the USA the causes seem to be plastics manufacture and use; suspicion has focused on plasticiser compounds, especially phthalates. Other androgen compounds include paper pulp production, PCBs, fungicides and pesticides, sewage contaminated with human oestrogen residue, mainly derived from oral contraceptives, and breakdown products of the pesticide DDT. The problem is manifest in developed and developing countries and needs to be carefully monitored.

Transmission of diseases

Infected humans can easily and quickly transport diseases to a new locality. Disease-carrying organisms are also easily spread via cargo, on vehicles, in ships ballast, as stowaways like rats and probably by migrating birds. Transport, wildlife, agricultural and public health authorities thus need to work together more closely, but in many countries government agencies do not co-operate. Once introduced, pathogens are often difficult or impossible to control or remove; the prevention of introductions is thus a good investment, but one that is neglected.

There are a number of cases where malarial mosquitoes have been stowaways on ships or aircraft and caused serious outbreaks. For example, a virulent strain of malaria spread from Africa to Brazil in the late 1930s. Increasing passenger air travel and the use of cargo containers on quite fast ships and aircraft have increased the risk. Before containers were developed, cargoes were loaded and unloaded at dockside which gave less chance for undiscovered stowaways such as rats and insects. Epidemic diseases have caused havoc throughout history, the last major outbreak being influenza in 1919. Today, haemorrhagic fevers like Ebola and influenza-type outbreaks like the recent sudden acute respiratory syndrome (SARS) cases threaten rich and poor countries.

Reducing the risk of disease transmission in food and commodities should be easier by means of effective inspection on despatch and at destination. Produce must be washed and disinfected, pasteurised, gamma ray irradiated, fungicide or pesticide treated. There are clearly trade-offs in treating food and commodities: treatment should not impose undue risk on the transporters or consumers, nor spoil the product, and it has to be cost effective. However, treatment potentially reduces huge costs of disease outbreak and lost labour.

Environmental managers can help advise on possible shifts in disease distribution and identify trigger events like El Niño/Southern Oscillation (ENSO) changes – the aim should be to help develop early warning and contingency planning.

Tourism-related pollution

Some countries host heavy densities of tourists and the resulting waste and sewage disposal may tax local authorities, or is devolved to individual hotels, which may result in unsatisfactory standards of treatment, monitoring difficulties and a dispersal of pollution. Labour may be lured from the land to work in the tourist sector and trigger neglect and environmental degradation in those rural areas. Water demands for visitors may be difficult to sustain, and recreational vehicles may disturb wildlife and fisheries. Tourists can introduce diseases to developing countries and convey infections to their own; these include sexually-transmitted diseases, mosquito-borne diseases and a wide range of viral and parasitic infections. The problem is how to make tourism as little damaging as possible, or even to make it pay toward environmental management.

The problem is that most tourist companies are based in developed countries, are large enough to negotiate vigorously, and are always able to threaten to take their custom, employment generation and foreign exchange earning elsewhere. Tourists easily object to what they see as high eco-taxes or too restrictive controls over their activities. But they can afford to pay reasonable levies and, if things are explained in a reasonable manner, generally do so. The precise definition of ecotourism is not established, but a reasonably acceptable measure of whether it deserves the title is to ask whether it reinvests significant amounts of tourism profit back into environmental management, as well as minimising environmental and social impacts (Mowforth and Munt, 2003). Unfortunately, green is often 'green-hype' and not as environmentally friendly as companies and host governments would like to make out. It is thus important for international agencies to monitor developments. Tourism development agencies can insist on eco-friendly building styles, aesthetic and environmentally sensitive locations, energy and water conservation, local sourcing of foods to ensure employment and retain foreign exchange, minimal use of chemicals on golf courses, and many other measures to reduce the eco-footprint. It is also desirable to ensure tourism development is sustainable. There is a growing literature on sustainable tourism, but much of it so far dealing with developed nations (Tribe et al., 2000).

There is the ever-present risk of a sudden change in tourist behaviour, triggered by almost anything – a scare over security, a recession or simply fashion shifts – any of which could catastrophically damage an industry built up at considerable cost. To try and spread the risks involved in tourism most countries seek a mix of types of tourism.

Better pollution avoidance and control

One key factor in avoiding and controlling pollution is information exchange: developing country administrators and citizenry need to know if there are risks associated with a new activity or compound. Then it must be established how safety can be maintained, and what should be done in event of an accident or misuse. This has been partially addressed by developed countries passing laws requiring better transfer of information overseas, including international hazard warning symbols and instructions, wherever possible in visual form and with multilingual text on secure weatherproof labels. There are also databases and call-in services which seek to provide information on products and activities – something much aided by the Internet. One such system is the Pesticide Pollution Network. NGOs have played an active role in developing these improvements, particularly consumer protection bodies, which are now co-ordinated by the International Organisation of Consumer Unions.

Several NGOs and institutions are very active and effective watchdogs for developing countries: for example, the Consumers Association Penang (Malaysia) and the Centre for Science and the Environment (India). Although, they might sometimes be accused of bias toward developed country interests, Northern-based NGO bodies like the International Institute for Environment and Development (IIED) and the World Resources Institute do seek to protect and assist the South. There are growing numbers of NGOs that vary in their outlook and independence, but which share a concern for environmental protection in developing countries; some conduct careful empirical study before campaigning, and others are 'ginger groups' which doggedly pursue causes, sometimes without checking facts. Such a mix is good, and the role of the media in developed and developing countries should not be overlooked. Most newspapers and TV stations now have 'green' journalists and react to threats and environmental problems and stimulate official and citizen action.

Environmental impact assessment (EIA) has spread from the USA since the mid 1970s and is now widely used by developing countries and by those funding projects. A major factor in the spread was that in 1975, four USA environmental NGOs sued the US Agency for International Development (USAID) for failure to do an EIA on some projects it funded in developing countries. By 1977 action had been taken to ensure all USAID-funded projects were preceded by an EIA; the 1978 US Foreign Assistance Act enforced this and soon most US and bodies elsewhere in developed countries had adopted EIA measures and published environmental management guidelines.

EIAs often provide poor results and are frequently side-stepped by developers. However, they are one way for citizens to get more access to information, a form of empowerment and participation, and a means for genuinely concerned developers to try to avoid problems. They also have the potential to improve the knowledge base about the impacts of activities without having to wait for an accident. Also, even when faulty, impact assessment does put pressure on planners and decision makers to take more care and look at research implications before they act – they are made to feel more accountable to the public and funding bodies.

It is difficult everywhere to establish cause and effect and liability for some pollutants. The impact of a harmful compound may vary greatly, according to the exposed person's genetics and state of health, the local environment, habits and many other factors. The manifestation of ill health may be delayed many years after exposure and it can be very difficult to prove it is due to pollution and not some other factor. Monitoring and better legislation may help, but proving chains of causation and making the polluter pay can be difficult. Perhaps for genuine reasons, possibly as an excuse, some companies cite a need to protect their trade secrets as a reason for withholding warnings and information on risks. In developing countries the insurance sector is weaker than in richer nations; this means there is less risk assessment and less pressure on developers. More insurance cover would mean the companies providing it would question risky developments, bodies and individuals would have more chance of funds after accidents to pursue claims, and that would propel developers toward more caution.

There are tricky legal problems. Where transnational companies are involved it can be difficult to hold their corporate officials responsible, so a manslaughter charge or damages claims are uncertain to succeed. Pursuing any liability claim across national boundaries is likely to be slow and costly (see the Bhopal case, Box 7.3). Where law and order are weak there may be groups using pollutants who feel there is little chance of being prosecuted, and if they are, often-limited fines are little discouragement compared with profits being made. In Amazonia small bands of gold miners (*garimpeiros*) pollute many rivers with mercury; there are alternatives, but these cost more and enforcement is difficult, so the problem continues.

Countries still vary a good deal in the standards they use. Without reasonably uniform global standards, monitoring, enforcement and compensation are hindered. However, many nations do now insist that multinational and joint-venture companies adhere to waste and pollution control rules which have developed since the mid 1980s. International bodies are also maintaining a better watch nowadays. There has been a move in developed countries away from dealing with pollution after its release – at 'end-of-pipe' – to avoiding emissions or dealing with potential pollutants before release. Progress in developing countries is limited by funding, lack of skilled personnel and equipment for monitoring and, because industrialisation has been rapid, less constrained by planning controls; sometimes, though, they can leapfrog and adopt new pollution control measures

There are still huge challenges when dealing with national and global common resources. The problem demands control of numerous traditional users, frequently with little or no documentation and sometimes a reluctance to invest in maintaining environmental quality. Development pressures can cause a degradation of traditional common resource management strategies; for example, states often sell exploitation licences or concessions to private enterprise, so traditional users are evicted. And the new user may get away with poor practices.

What can be done to improve pollution management? There are a number of measures:

- A workable strategy for managing (ideally avoiding) pollution: each country and international bodies should develop one, and it should look sufficiently far ahead to ensure there is room for manoeuvre. Unfortunately, few countries have satisfactory long-term strategies and so many stumble from reaction to reaction, rather than developing proactive planning. What progress has been made can be judged from the World Resources Institute (1996) directory and bibliography of country environmental studies, environmental strategies and pollution measures.
- Integration: as far as possible, pollution management should be integrated with other fields of environmental management; some go a little further and advocate a holistic approach.
- Risk assessment: at an early stage in planning, the potential dangers should be predicted and presented to developers, administrators, engineers and possibly the public.
- Impact assessment: at the outset of planning, early enough to allow consideration of alternative means or sites, an assessment should be made seeking to document the negative and positive impacts of the development. The results should be presented to the developers and authorities as a statement. In developed countries the statement would also be released to the public; however, in developing countries there may be problems in this because of widely scattered populations, limited literacy, poor people with little time to consult documents, governance with little history of consulting the people and risks of civil unrest or speculation. Potentially, impact assessment can encourage or force the developers to think carefully; it can also allow the public, or at least NGOs representing them, to participate more effectively in planning. The problem is the cost and the need for expertise and, crucially, effective legislation and planning measures to ensure the assessments are not corrupted or ignored.
- Health and safety at work: ideally authorities, possibly unions and NGOs, monitor conditions and ensure risks are minimised for employees and probably nearby bystanders. However, a number of developing countries attract industrial development in part as a consequence of limited unionisation and a 'docile' workforce. Consequently, NGOs and government monitoring is often all there is.
- Pollution inspection and legislation: most developing countries have regulations and laws in place, but these may lag behind need (not surprising given the speed of development and the sheer diversity of new technology), and resources to monitor and enforce are limited. The approach was applied early on in developed countries to try and control pollution, often with good effect and on limited budgets, so it should have promise in developing countries, but in reality enforcement is often poor.
- Modern pollution monitoring equipment: this is becoming more sensitive and robust, capable of automatic remote functioning, and less expensive, so

assessing pollution patterns and identifying polluters is less of a problem. When fines are imposed these may be lower than those in developed nations and the companies simply see it as a bearable cost.

- Consumer protection: while mainly aiming at protecting consumers, bodies active in this field have also been a significant force for general environmental management improvements in developing countries. Indigenous and expatriate NGO groups or institutes have appeared in many developing countries since the 1980s, usually relying on media support and public demonstration. These bodies inform and warn about, and object to, pollution issues. NGOs may call for or conduct risk assessments or impact assessments. Increasingly these bodies have 'teeth' and can lobby and pressure governments and publicise issues worldwide. The latter has been much aided by the establishment of e-mail and the Internet – today it is becoming impossible for serious pollution and other environmental problems to be deliberately hidden from global scrutiny. Some of these bodies have undertaken to stand up for developing country interests in international pollution debates: for example, challenging statistics on global warming published by USA sources

- Commercial controls: a growing number of consumers are worried about contamination with pollutants, or object to loss of biodiversity caused by pollution. These consumers are willing to lobby for controls, pay for measures and, in particular, refuse to buy goods that do not meet their demands. Organic produce commands a premium in Western supermarkets and there is a growing lobby for ethically-produced food and commodities. There is now consumer demand for fair-trade beverages, where smallholder producers are paid adequate prices and given other supports by the marketing company, wood from managed forests and other 'eco-friendly' products. There is a growing global trend toward encouraging ethical business, and part of that is reduction of pollution.

- Trade and market forces: world trade is increasingly governed by international rules which, in theory, should make it easier to enforce monitoring and control of pollution (Chander and Khan, 2001; Low, 1992). There are also ways for developed country governments, companies or organisations to make agreements with developing countries to control pollution: for example, debt-for-environment swaps, waste sorting for recycling using cheap labour in a developing country and tradable emissions controls.

- Anti-pollution foreign aid: given that pollution crosses borders, it makes sense to provide aid to combat pollution. If developed countries wish India and China to adopt pollution avoidance, or install expensive carbon emissions controls or chlorofluorocarbon (CFC) emission reduction measures, they will have to ease the cost by targeting aid that way.

Disposal of dangerous compounds

Particularly dangerous compounds can be treated in incinerators, thereby hopefully avoiding environmental impact. Incineration is not foolproof or cheap, but it is better than dumping dangerous waste. Incinerators have to maintain high enough temperatures – for PCBs, over 1,200°C for more than 60 seconds – possibly for large quantities of pollutants. This demands costly equipment, good management, skilled maintenance and constant monitoring of emissions for dangerous fallout. Poor countries may not be able to afford this and will have to export the waste for treatment, either in a developed country or in a facility shared by a number of developing countries. Ship-based incinerators are available which can visit countries lacking their own incinerators at regular intervals, and in North America incinerators have been mounted on road and rail trailers. However, adequate supervision of portable incinerators may be difficult. For some pollutants, bioremediation may be a good option, if safe and effective micro-organisms can be found and cultured. Such treatment might be in sealed digesters at a special plant, or *in situ* where soil and groundwater have been extensively contaminated. Currently techniques are being developed and will probably be commercially available within five to 10 years. Bioremediation might cope with even the most challenging toxic compounds: reports in 2003 claimed a bacterium capable of digesting dioxin had been identified in Germany and might be improved for practical bioremediation (Graham-Rowe, 2003).

Whether incineration or bioremediation, the technology is likely to be developed by business or a developed country, and will need to be purchased by poorer nations – pollution control is thus likely to mean dependency.

Concluding points

- Sustainable development, whether in rural or urban environments, demands effective handling of outputs – waste and pollution – as well as inputs. Sustainable urban development is emerging as a goal but actual achievements are still limited.
- Water supply, refuse and sewage disposal are pressing problems. Until these are improved, human well being suffers. In the coming decade a lot of attention will focus on these challenges. Energy, transport and housing difficulties are the next priority. Cities will have to cope with rising populations and the possibility of global environmental change.
- Pollution management demands the establishment of ethics as well as regulations, monitoring and enforcement. There is an ongoing shift to a proactive approach and adoption of the polluter-pays principle.
- Many pollution and waste management issues are transboundary, and some are global. Solutions and controls demand co-operation and funding from both developed and developing countries.

Further reading

Farmer, A. (1997) *Managing Environmental Pollution.* Routledge, London. [Readable introduction to pollution issues.]

World Resources Institute (1997) *World Resources 1997: the urban environment.* World Resources Institute, Washington (DC). [Coverage of urban environmental issues by a leading environmental NGO – available on the web, see below.]

Environment and Urbanisation, special issue October 1992. [Journal, special issue on sustainable cities.]

Drakakis-Smith, D.W. (1987) *The Third World City.* Methuen, London (2nd edn. *Third World Cities*, 2000. Routledge, London). [Good introduction to developing country cities and towns and urban policy; Chapter 4 reviews environmental issues.]

Hardoy, J.E., Mitlin, D. and Satterthwaite, D. (1992) *Environmental Problems in Third World Cities.* Earthscan, London. [Review of urban environmental problems in developing countries.]

Websites

Urban Studies, journal, Carfax Publishing: http://www.journalsonline.tandf.co.uk/app/home/issue.asp?wasp (accessed March 2004)

World Resources Institute – downloadable form of the reference listed earlier: http://www.wri.org/wri/wr-96-97focful.html (accessed January 2004)

Global Development Research Center virtual library – environmental management: http://www.gdrc.org/uem/how-tos.html (accessed January 2004)

8 Environmental threats

Key chapter points

- This chapter examines threats, risks and hazards, vulnerability, adaptability and resilience. Threats include natural random and more predictable physical phenomena; human-induced physical changes like global warming, pollution and introduction of alien species to new environments; social problems such as warfare, reduction of social capital and so on; economic problems ranging from depressions to trade regulations; technology innovation problems; energy supply problems; population growth; and ageing.
- Environmental management should alert developers to significant threats, limits and opportunities. Wherever possible environmental management should seek to orchestrate suitable solutions.
- Alerting decision makers and the public to threats is only part of what is needed for threat reduction. People may ignore threats for many reasons; in developing countries poverty often limits people's options for avoidance and mitigation.
- The poor tend to be more vulnerable to threats. Also, development over the last hundred years has made developing and developed country populations more vulnerable and probably less adaptable. In the developing countries breakdown of coping strategies has been a factor. Everywhere growing population and increased consumption stresses the environment and society and makes disaster recovery more difficult.
- History and palaeoecology can be useful to forewarn of risks and suggest how to construct contingency plans.

Before modern times, future events were largely seen to be in the hands of the gods or God; nevertheless, settlements were usually sited to minimise known recurrent hazards such as avalanche, flood, piracy and so on. People have always been aware of threats. By Victorian times financial institutions in Western countries were methodically reviewing risks More recently many nations have established civil defence arrangements to help cope with threats. Since the early 1980s environmental management has been developing risk assessment and management techniques.

Hazards and risks, vulnerability, resistance and resilience

A hazard is a potential threat; a threat is when there is reasonably clear danger. A risk is the probability that a hazard will happen or, more precisely, the likelihood that an event will coincide with elements which can be affected. There is some variation in definition and measurement of risk between disciplines: engineers, toxicologists, economists and ecologists all differ. One definition of risk is:

Hazard × vulnerability

However, an engineer might see it as:

probability × consequence

A toxicologist would probably define it in terms of:

threat > pathway > target characteristics

Some see risk in purely human well-being terms, others are also concerned for the environment. Risk management can be conducted at various times; often it is undertaken when a problem is manifest; there are also good arguments for it to be conducted in advance of any development, and for it to be repeated at regular intervals. Risk management can be subdivided into risk identification; risk assessment; risk perception assessment; risk communication; risk avoidance, mitigation or control; and emergency responses.

A risk assessment identifies, and measures, the significance and likelihood that a threat will happen. It is also likely to involve assessment of the vulnerability of humans, ecosystem or environmental features. Vulnerability can be defined as the degree of sensitivity to an impact; it need not imply awareness that the threat exists. Alternatively, it can be defined as the characteristics of a person or group in terms of their capacity to anticipate, cope with, resist and recover from the impact of a natural hazard (Blaikie et al., 1994: 9). Broadly, resistance is the ability to withstand a hazard; resilience is the ability to recover from it. Although there can be no precise definition, a disaster may be described as the realisation of a hazard with serious consequences (Whittow, 1980), or the serious disruption of everyday patterns of life (Wijkman and Timberlake, 1984). Some authorities have tried to apply human or monetary benchmarks: for example, defining a disaster as when 'there are more than 100 dead'. Like threats, disasters can range from a very brief event like an explosion to something, like soil degradation, stretching over centuries or more.

Some risks are voluntary and some involuntary; sometimes threats are linked to opportunities. The situation is clarified by the following example: someone wishing to cross the Atlantic may seek to do so by rowing boat or ocean liner. The hazard is the same – a risk of drowning. However, the risk is much higher in the rowing boat.

The impact depends on the nature of the threat, on vulnerability and recovery, and is affected by resilience and availability of aid from unaffected areas. Many of the known hazards have been assigned scales of magnitude: for earthquakes the Richter and modified Mercalli scales, for winds the Beaufort scale, and so on. So a listing of the recurrence of disasters can be judged against benchmarks. For the public the most meaningful magnitude benchmarks are numbers killed or injured and cost (see Whittow, 1980: 40–50 for some of these scales).

Human activities can increase vulnerability to threats and may also alter their natural occurrence and generate 'man-made' hazards: for example, deforestation is likely to alter the runoff from a catchment, often increasing the risks of floods and silt deposition. Also, vegetational changes like forest clearance may trigger human disease outbreaks as people come into closer contact with wildlife reservoirs of infection and vector insects (see Figure 8.1). Threats may be natural, human-induced or often a complex mixture of both, made even more difficult to unravel by cumulative effects (more than one impact interacting) and by positive or negative feedbacks (processes which exaggerate or counteract a trend). Similar events at different points in time may set in motion different chains of causation; what happens once need not produce the same event the next time, so environmental managers must be prepared for surprises (Kates and Clark, 1996).

Figure 8.1 *Forest clearance (South Asia). There can be considerable off-site impacts, such as silting of streams and outbreaks of human disease, as people come into contact with forest insects.*

Source: Author, 1978.

People differ in their perception of threats: they may range from disinterested or complacent, even in the face of significant risk of major hazard, to alert and reactive. So there are a number of possible reactions to a threat: it may be ignored; efforts might be made to avoid or mitigate it; or contingency plans and relief measures could be readied to help recovery (Varley, 1994; Anderson and Woodrow, 1998). Risk assessment and management deals with trade-offs: economic and political versus technological capability. Risk assessment is to some degree subjective and people may ignore warnings and judge things 'acceptable'; typically more risk will be taken if there is a benefit, especially a profit, to be made. Benchmarks used to assess risk include established standards like toxicity tests seldom made on humans, failures per kilometre travelled, and hindsight.

Environmental management must warn of threats and environmental limits which should not be exceeded, map the dangers and degree of vulnerability, identify development opportunities and suggest appropriate responses. Effective early warning demands the identification of potential threats and monitoring to spot if any relevant critical thresholds are being approached. It is now relatively easy to obtain maps and geographical information system (GIS) readouts of vulnerability to and threat of landslide, flood, avalanche, tsunami, drought and so on.

Warfare and civil unrest threaten environmental management and hinder sustainable development efforts in many countries – and problems may continue long after actual hostilities cease because people have been driven from their land, and weaponry like landmines may remain active. Chemical compounds like defoliant may pose an 'environmental' problem long after fighting has ceased; this was apparent after the Vietnam War, during which large areas of defoliated forest changed to rank grassland and never recovered because of soil and groundwater alterations caused by tree loss. Other areas of Vietnam were replanted with robusta coffee and are currently helping to drive down market prices and ruin many of the world's growers. Unrest in many parts of the world, plus the impact of restrictions to control pollution and rising costs of pesticides, has meant the neglect of insect control. Consequently, locust damage and mosquito breeding have made a comeback in some developing countries since the 1970s – Mauritania announced a growing locust problem in 2004.

The invasion of new areas by organisms which out-compete the established species can be natural because some physical change, biological adaptation or accident of nature aids dispersal, but is today more likely to be assisted by humans through tourism, business transport and hobby collectors of plants and animals. Charles Elton warned of the threat in the 1950s and new examples increasingly occur (Elton, 1958). Sometimes the invasions pose a threat in the form of new or increased outbreaks of human disease, or they may threaten food production (Food and Agricultural Organisation [FAO], 2001b); infrastructure through damage to channels and banks by burrowing crustaceans, molluscs or rodents; choking of pipes, canals and waterbodies by weeds, molluscs or algae; or they may pose a nuisance. Many developing countries are very dependent on one or a few crops for export earnings and are thus vulnerable to introduced organisms. The move toward free trade means that effective plant and animal quarantine measures must be acceptable to General Agreement on Tariffs and Trade/World Trade Organisation (GATT/WTO) authorities and monitoring must be improved.

The bovine spongiform encephalopathy (BSE) and foot-and-mouth livestock health problems in the UK recently illustrate how easily outbreaks can occur and how costly they can be. The causative organism of foot-and-mouth had probably been introduced by migratory birds or the importation of uncertified meat. Warning of a threat may not be enough – the UK has long been aware of the dangers of foot-and-mouth – the problem is getting people and administrators to act in an appropriate manner to reduce the risks, and global monitoring to get early warnings. In the nineteenth century Europe nearly lost all its grapevines to introduced insect pests that transmitted disease. Cocoa producers are vulnerable to fungus diseases, and many other developing country crops are at risk.

Some threats, which cannot be prevented, might be controlled or induced in a way to cut damage compared with their unhindered effect; for example an avalanche can be deliberately set off before too much snow accumulates and at a time when people are evacuated. There are hopes that if the stresses causing earthquakes can be seen building up an area might be evacuated and then some means used to trigger it. Similar pre-emptive controls might be possible for dangerous gas emissions from lakes, and are already used against the threat of forest and bush fires and landslides.

Foreseeing threats in advance is often difficult; some 'creep up' in an insidious way – one moment there is no problem, and the next a full-blown disaster. The latter situation can give rise to a 'frog-in-a-kettle' scenario, where people experiencing gradually deteriorating conditions – for example, gradual soil erosion – may fail to act because there is no clear crisis point. A frog gradually warmed reputedly stays and boils, whilst one suddenly placed in hot water will try to leap free. Some threats are sudden and obvious, some slow and obvious, some slow and insidious, others sudden and insidious – the latter two are the least likely to provoke a response. An example of a relatively slow and insidious threat is human immunodeficiency virus/ acquired immune deficiency syndrome (HIV/AIDS); an example of a sudden and obvious threat is the 1918–19 'Spanish' influenza pandemic. The latter is little remembered today but it killed around 22 million people – more than First World War military casualties.

Some very rare events can have catastrophic impacts, yet without historical experience people are unlikely to treat the threat seriously. Response depends largely on human perceptions of severity, extent and frequency of recurrence. Individuals, state agencies, media and non-governmental organisation (NGOs) may be able to educate people to respond to threats and can support monitoring and early-warning activities. For example, the FAO established the Global Information and Early Warning System (GIEWS) in 1971 to provide early warning if world food supplies looked likely to become problematic. Japan, Hawaii and Pacific-coast USA have tsunami early-warning systems, and a number of countries have avalanche-alert systems. Weather forecasting has improved significantly in recent years, allowing improved hurricane warning, and river basin managers can make use of it to prepare for floods. Earthquake early warning is not yet developed to a useful degree and the same can be said for volcanic eruptions, although there are promising signs the latter might be developed.

Further complications can be identified. Different groups of people vary in their response to a threat: e.g. locals, regional administrators, state administrators, international agencies, world opinion, different age groups, gender groups, class groups, ethnic groups and others. Also, each may alter its response over time; consequently, people often react differently to successive similar threats. Some people are able to adapt, some panic, some are fatalistic or lazy and many may have too limited resources to make much response. Access to opportunities to overcome threats is crucial, and some groups have far more access than others, usually the rich. Where a single disaster may be withstood, repeated impacts of the same sort or an unfortunate combination of events may spell catastrophe even for a resilient group of people or organisms.

Poverty usually means people are condemned to situations in which they have little in the way of reserves to resist threats or to recover from them, and 'living near the edge' in such situations they are often very vulnerable. Administrators and politicians may often not respond to a threat because they are not well-enough informed; but sometimes it is because there are no funds, or action does not seem likely to win them promotion or votes. Forecasters and administrators run the risk of being seen to be 'crying wolf' – if they act proactively and then find the threat has not materialised, they lose face or worse and will be reluctant to act next time. So if any action is taken the strategy increasingly adopted is to seek a win-win solution. This is a solution that has benefits if the threat becomes reality, and also offers something if it does not – hence it is usually a compromise, and not necessarily the most effective route. Another common response is to wait and see or at least delay to seek further proof and/or wait for technology to improve. Few modern administrators or politicians look further into the future than about five years; they generally show little concern for something 'not in my term of office'. Unfortunately, many resource development and environmental management issues demand a much longer planning horizon. There is another unhelpful response that is quite common in developing countries and in some richer nations, which is for those active in environmental management or responsible for human welfare to have to report to headquarters and await permission to act. The delay may be long, and there may be reluctance to bring 'problems' to the attention of superiors for fear they are seen to be troublemakers, overzealous or unable to cope. This top–down administration may have other impacts. It means centralised control which is likely to be out of touch; those with the education and skills to make wise responses may resist postings to remote sites and, understandably, in an uncomfortable and possibly dangerous environment, be reluctant to 'get their boots muddy'. There is a further possible problem: the developing country specialist may have been trained in a developed nation, and so is not sufficiently sensitive to issues. Where expatriate expertise is being used the same risk is present, coupled with the likelihood that they will move on before problems manifest without training locals. The Internet may help resolve these problems; as distance learning becomes available more developing country staff can be educated at a reduced cost. However, an adequate fieldwork and practical component may be difficult to deliver with it.

One reaction to a threat is to insure against it. Insurance spreads the risk; a large number of premium payers contribute manageable sums and, provided not too many make claims at one time they should get adequate compensation. Unfortunately, some threats materialise into problems, which affect many people at once, so that spreading the risk by insurance fails. Insurance companies may counter by reducing payouts; re-insuring, so that the burden is spread, possibly worldwide; inserting exclusion clauses into cover; imposing excess charges; refusing high risk clients – but for some threats such measures cannot resolve things. Some insurance companies have great experience and have developed very effective risk assessment approaches, which can assist environmental management. In the UK, for instance, commercial flood risk maps have recently been made available to environmental planners (McCall et al., 1991; Smith, 1992; Hewitt, 1997; Alexander, 2000).

Once a threat is realised, it may be possible to retrofit infrastructure or modify ongoing developments to reduce impacts and vulnerability. Adaptation to environmental and socio-economic threats is both a cultural (learning) and in the longer term a biological process (Anderson, 1968; Hewitt, 1983; Oliver-Smith, 1986; Adger, et al., 2001). So education may reduce vulnerability and help prevent some problems. There may be situations where a disaster offers opportunities: depopulation through epidemic may enable land reforms or solve employment difficulties and prompt longer-term development, as seems to have been the case after the 13th century Black Death in Europe; without the upheaval, social, cultural and economic change would probably have been slower. Some disasters causing serious damage in the short term may offer opportunities over the longer term; for example, some volcanic ash-falls, which destroy crops and injure people, may weather to yield fertile soil. Conversely, disasters may trigger disruptive adaptations; for example, eco-refugees may spread and cause secondary impacts in a number of places.

Recently a lot of attention has focused on how humans affect the environment, especially through global warming; much less effort has been devoted to assessing how the environment affects humans. During the last few decades knowledge generated by palaeoecologists and environmental historians, plus the stimulus of some recent severe ENSO events like that of 1997–8, have helped stimulate interest in environment–human interrelations (see Fagan, 2000; Davis, 2001). Keys (1999) tries to convince his readers that there have been strong links between environmental events and human fortunes, singling out AD 535–6 as a global crisis point when a volcanic disaster triggered a chain of adverse weather, hunger and plagues. However, while neo-environmental determinism makes fascinating reading and may offer valuable warnings, a good deal of it is not sufficiently thorough and objective. A historical correlation between, say, a volcanic eruption, evidence of failed harvests and social calamity, does not prove causation. There are very few indisputable correlations between environmental events and human fortunes, the AD 79 Vesuvius eruption being one; many others remain in dispute because some evidence suggests environmental and other socio-economic causation (Barrow, 2003: 65–8; Crumley, 1994).

If natural disasters can be *reliably* proven to be periodic or quasi-periodic through research by environmental historians, synoptic climatologists and palaeoecologists, then administrators will take the threat seriously and useful forecasting is possible. That situation seems to be approaching with respect to ENSO events.

Thirty years ago, Meadows et al. (1972) fed information on what they saw as critical development issues (food supply, pollution, population growth and so on) into a computer systems model; the result was a prediction of future trends which showed crucial limits being exceeded. Their sequel (Meadows et al., 1992) re-examined the predictions and concluded a crisis would still happen within a generation or so unless environment and development was better controlled; recently others have explored 'future development crisis', one predicting serious trouble as soon as 2030 (Mason, 2003).

Global environmental change

Considerable resources have been devoted to predicting the likely impacts of global environmental change, which is most widely assumed to be human-caused warming – the 'greenhouse effect'. A number of university and international institutes have modelled its probable impact on agriculture, human and livestock health, and other important issues (see Intergovernmental Panel on Climate Change [IPCC] available online at http://www.ipcc.ch, accessed February 2004). Global environmental change is multifaceted; there is ongoing natural change – random, quasi-periodic and periodic – and there is that partly or wholly caused by humans. Examples of natural change include the solar cycle, geomagnetism, orbital attitude (all possibly periodic); and volcanic and seismic events, asteroid and comet earth impacts, natural species extinction and evolution (no obviously repetitive pattern established). Superimposed on natural changes are those caused by human activity: pollution, land clearance, hunting and so on. So far most human impact has been through hunting and pollution, but transportation is increasing the movement of organisms, and genetic engineering may become major causes of change. Some argue that human influences are now so great that a new current geological period should be recognised – the Anthropocene.

Greenhouse warming is presently seen as a 'key' environmental issue by many people and it enjoys huge media coverage. Developed and developing countries have been trying to assess likely impacts and predict future scenarios. Will richer nations be better able to withstand changes? Will the poorer be at a disadvantage, or will it be vice versa? Where will the greatest problems arise? Can developed and developing countries reach agreements to mitigate, avoid or adapt to threats?

The last 20 years have seen much-improved collection and sharing of data. While not by any means satisfactory there has been some progress in establishing talks and drafting agreements. Some states have begun to co-operate in lobbying for pollution controls and environmental change-related foreign aid. Assuming that the way global warming will proceed is uncertain, that the impacts are not easy to accurately predict, that change is already underway, and that pollution control agreements are relatively ineffectual – one could argue that environmental managers could achieve more by focusing on how to reduce people's vulnerability and improve adaptability. Adaptability seems to have decreased in recent decades in rich and poor countries. Improving adaptability should be relatively easy compared with cutting global pollution, and it would serve against a wide range of unexpected natural and human disasters and environmental change.

Environmental change may well have urban health impacts, but warnings are difficult to assess because many complex factors are often involved. Predictions so far are too often based on guesswork: for example, global warming is widely assumed to be likely to cause a spread and increased incidence of malaria. Yet in Europe, between the mid fourteenth century AD and the 1840s, malaria was common, although conditions were significantly cooler then than now. Factors other than climate seem to explain the demise of the infection in Europe.

One of the more likely impacts of global warming could be shifts in energy use and consumption of goods as rich and poor countries implement economic controls and introduce new technology to honour agreements over carbon emissions and adapt to climatic change. Environmental management must forecast and prepare for changes, and it must try to do so as objectively as possible.

Global environmental change is a transboundary issue – its causes are shared by all countries and its consequences will be faced by the whole world. In the last four decades there has been much more awareness of transboundary issues and the international impact of problems that would in the past have been mainly regarded as affecting only the poor of developing countries, e.g. multi-drug resistant tuberculosis.

Greenhouse warming could have the opposite impact on some parts of the world; warmer tropics might mean more water vapour and greater snowfalls at the poles, cooling higher latitudes. Another risk is that warming will alter thermo-haline ocean circulation and in the Atlantic this might 'switch-off' the Gulf Stream and cause a sudden deterioration of Western European and possibly eastern USA climatic conditions. However, this has had little attention; most assume the future will be one of steady, manageable and predictable warming. Environmental managers concerned with development have to be aware of people's perception of threats, which may not be logical or predictable, and must weigh human vulnerability against threats they identify. The currently dominant assumption or wishful thinking, that global warming will proceed in a gradual and steady manner, is unwise. Palaeoecology shows that there have been sudden thermo-haline circulation shifts in the Atlantic during the post-glacial period – the last 13,000 years or so – leading to sudden cold phases in Europe. A repeat would spell disaster to world food production, economic development and possibly peace. A wit once remarked that 'scientists and artists both tend to fall in love with their models'; many of the threats faced are very difficult or impossible to predict sufficiently accurately to give useful early warning, yet a lot of effort and funds are being spent on modelling. A better response would be to focus on improving food reserves and dispersing and diversifying food production, so that at least people could be fed long enough to have a chance to adapt. Commerce and globalisation may not favour such precautions; if profit drives much of the world's agriculture, there is little incentive to stockpile food reserves or diversify small-scale farming strategies.

Are modern humans more vulnerable?

There has been a realisation in some quarters that the strengths which have favoured human survival in the past – adaptability and a willingness to think ahead – may have diminished (see Box 8.1) (Barrow, 2003: 100–22). Many would argue that the last 5,000 years have been unusually stable in terms of environmental change, but with global climate change triggered by humans, can that last? Nowadays, people are generally far less adaptable and mobile than once would have been the case: populations have grown, basic survival skills have often been lost and, especially in developed countries, a cushioned lifestyle has made many less robust and resilient.

Box 8.1

Factors that have increased human vulnerability

Increasing dependency: services, food production and many other things are increasingly supplied from outside a community. A problem with a supplier can have worldwide impact.

Biodiversity damage: this is reducing options for maintaining agriculture, pharmaceuticals production and so on. Human activities have been depleting natural biodiversity, and the range of traditional crop and livestock varieties used.

Global (anthropogenic) environmental change: caused by pollution and vegetation clearance (warming, acidification, environmental pollution and stratospheric ozone loss). There is more likelihood of change.

A decreased chance of moving to adapt to change: the world is crowded and people are used to living in one place. Hunter-gatherers and generalists can move relatively quickly and easily.

Terms of trade and agricultural development strategies: that may discourage growing and storing sufficient food reserves and the production of crops subject to market fluctuations.

Communication, settlement, social and healthcare developments: these increase risk of disease due to rapid travel, crowded cities, misuse of antibiotics and the concentration of migrants and refugees who are weakened and vulnerable.

Weapons of mass destruction for armed forces and terrorism: potential to dislocate and debilitate populations and disrupt the environment.

Increasing human population: puts stress upon resources and environment and there are more to feed if there is a disaster.

Intensification of agriculture, industrial and medical use of biotechnology: risk of new 'out-of-control' challenges.

Populations who have little 'survival talent': these people would find adaptation difficult.

Complex technology and livelihood strategies: these are easy to disrupt and difficult to service, repair or replace. People are unlikely to have skills for repair or maintenance.

HIV/AIDS (SIDA): already causing serious problems in some regions. Livelihoods are breaking down and environmental degradation and increased poverty follow.

Source: Based on Barrow, 2003: 102 (Box 4.1).

Globalisation and rapid travel have left both rich and poor countries in many ways more vulnerable to disasters than in the past: diseases can now rapidly spread across the world carried by air travellers, and if specialised manufacture for a global market is disrupted the implications can be widespread, e.g. a world shortage of microchips followed the 1999 Kobe Earthquake. In developing countries, rising populations, environmental degradation and erosion of social capital and traditional survival strategies have often increased vulnerability to threats. Modern humans in rich and poor countries are less secure from environmental threats than they like to think and nature is not as benign and stable as most hope (Bankoff, 2001; Barrow, 2003: 102). Environmental management must assist governments to improve their citizens' adaptability and reduce vulnerability. Unfortunately, many anticipatory actions, and forward planning looking more than a few years ahead, get little support. Spending on improving adaptability and reducing vulnerability to a wide range of threats is wiser than expenditure on global warming controls.

Threats that recur, but only over hundreds or thousands of years, but which could be catastrophic, are ignored. This is unwise but understandable; people are reluctant to address threats that are seen to affect others at a distance in space or time.

In spite of recent improvements in forecasting extreme environmental conditions, statistics indicate that growing numbers are killed by natural disasters in developing countries. This is in part due to population increase and settlement of more risk-prone areas as land becomes scarce, and also because poverty generally increases vulnerability. The mid-twentieth to twenty-first century has witnessed the rapid growth of cities, especially in developing countries; large cities are complex and often vulnerable ecosystems, and once disrupted may recover with difficulty.

Today's development projects, programmes and policies are often cumbersome and inflexible, and any change in human or environmental circumstances may cause considerable trouble. For example, a large irrigation scheme may be constructed with little leeway for coping with altered runoff drainage in order to save funds, consequently even a small increase in storm intensity or alteration of crops or farming methods may cause problems. Technology may give new options but may also lock people into activities that pose a threat. Technology should be reviewed by environmental managers to try to ensure that it reduces human vulnerability. There may be situations where a threat demands a totalitarian response – faced with sudden and unexpected serious challenges, environmental management will have to be flexible (Eisenberg, 1998: 299).

Environment–trade problems

Over the last 30 years or so a number of regional free trade agreements have been established, including the North American Free Trade Agreement, the European Union and the Association of South East Asian Nations. The principles of GATT were extended by the WTO in the 1990s, so that by the start of the millennium free trade rules determined by the WTO were effectively in force globally. GATT and the WTO had by 1994 laid down that there should be no discrimination or trade barriers. Ensuring that GATT does not cause environmental problems is the concern of the Committee on Trade and Environment of the WTO.

One WTO problem has been with the application of the precautionary principle. Because this is difficult to pin down legally, there has so far not been a full consensus on a definition, or whether or not to formally accept it. Consequently, there have been a number of situations where problems have arisen; for example, the WTO objected when Europe wished to restrict the import of growth hormone-fed beef, mainly from the USA. The WTO viewed this as a trade restriction because the precautionary principle was seen to be invalid by virtue of the threat being insufficiently proven (Robinson and Kellow, 2001: 21). A number of environmental management agreements have sought to use the precautionary principle; for example, the UN Biodiversity Convention and the Rio Declaration on Environment and Development hold that, even if full evidence is not available, where there is a serious threat, measures should be taken to minimise it (for the EC view see European Commission, 1998 and the EC website: http://www.europa.eu.int/, accessed February 2004). GATT Article XX (b) does allow measures such as quarantine for protecting human, animal or plant well being; but in practice the issue of justifiable risk is still problematic. Negotiations at Cancun in 2003 do not seem to have advanced things much.

There are international agreements that do reasonably effectively embrace the precautionary principle and try to fit in with WTO demands; the Kyoto Protocol of the Framework Convention on Climate Change, Article 2:3, for example, states that policies and measures seeking to reduce climate problems should be implemented in such ways as to minimise any adverse effects on international trade.

Opportunities for disagreement over trade and environment issues, especially relating to the precautionary principle, still abound. One example of trade barriers versus environmental protection would be where a state, unable to tell tins of tuna caught by dolphin-friendly methods from those containing fish landed in less environmentally welcome ways, imposes an import control on all tinned tuna. While this would help protect marine mammals it would probably infringe GATT rules. Some free trade measures conflict with health measures, like quarantine; trade in hazardous materials banned in some nations but not in others could hinder measures aimed at protecting the environment, like the Convention on International Trade in Endangered Species (CITES) (Von Moltke, 1996).

Since the early 1990s most of the former communist states have embraced some form of free-market economy. These transitional economies have shifted from centralised governmental control toward increasing activity by commerce. It is still not clear how effective environmental management will grow out of the confusion (Pryde, 1991; Sims, 1999; Rojsek, 2001). Will the companies take the initiative? Will the state have to impose controls, or can there be a joint venture approach?

Energy supply threats

Compared with the 1960s and 1970s, worries about future energy supplies have been relegated to a side issue. There is ongoing research and development into clean sustainable energy sources; some would meet developing country needs, in particular appropriate and accessible technology, which is practical and affordable.

Developing countries have been widely pressured or tempted to develop hydro-electric power to avoid petroleum imports and to keep down carbon emissions. Given the numerous, often negative, socio-economic and environmental impacts, hydro-electricity generation is not as green as some try to make out. Worse, large amounts of methane are often emitted from large reservoirs and the irrigation development associated with some large dams is also likely to raise methane emissions – not be as green in terms of avoiding global warming as many think. Large dams are also prone to silting up, so are sometimes not the sustainable energy sources that they are made out to be.

A number of developing countries are effectively in a crisis situation with respect to fuel wood. The supply of fuel wood and charcoal can be major causes of environmental degradation, and the costs of purchasing the fuel can absorb a large amount of poor people's income. Finding alternative energy sources for millions of poor people scattered over wide areas is not easy. Providing sustainable supplies of fuel wood may be possible in some areas, through woodlots and hedgerows. Care must be taken to ensure the fuel-wood supply measures do not destroy biodiversity. Carbon emissions from burnt fuel wood would be at least part-balanced by woodlot regrowth.

In urban areas, sewage and refuse offer some opportunities for biogas generation, but smaller settlements may simply not have suitable waste or the resources to treat it. Solar power via photovoltaic panels, local wind turbines and mini- hydro plants, which generate less than 6 MW, could support village lighting, TV and small refrigeration units for healthcare or veterinary use. The challenge remains to find replacement fuels for cooking and heating.

Ageing

In developed and developing countries, age is becoming a matter of concern, especially in poor countries where there is unlikely to be much state or private provision for pensions. The proportion of elderly people surviving in developing countries is growing (Wright, 2000). In some countries the cost of care may compete with environmental management for funds. However, this group could also be a valuable resource which should be tapped to support environmental management; they might be used to help staff various initiatives, to educate children in green issues, or simply provide voting support if recruited to the cause.

HIV/AIDS

Disease transmission problems were covered in Chapter 7; HIV/AIDS is considered here because in some rural areas there is a growing infection rate which has or will affect labour availability. Labour shortage may result in breakdown of established agriculture and cause environmental degradation. Large numbers of families left without one or both wage earners will become an economic burden as well as a source of great misery. There is also a possibility that funds will be diverted from environmental issues to cope with the welfare problems.

Concluding points

- If ongoing threats can be identified and early-warning signs monitored, there is a chance of avoiding some problems, mitigating measures will be better prepared, and there is more opportunity to adapt.
- Some risks occur at random or result from causation too difficult to unravel. There will continue to be unexpected threats no matter how good environmental management may be.
- Modern humans may be more vulnerable than most people realise. Efforts need to be made to improve adaptability and resilience.
- There is growing interest in environment–human interactions. However, attention and expenditure are mainly on global environmental change – specifically 'global warming'. Outcomes are uncertain and controls difficult. It would probably be better to spend on generally reducing human vulnerability, which would serve against any threat. Developments could also be made more flexible and adaptable, rather than seeking to minimise costs.

Further Reading

Blaikie, P.M., Cannon, T., Davis, I. and Wisner, B. (eds) (1994) *At Risk: natural hazards, people's vulnerability and disasters*. Routledge, London. [Thorough exploration of risk and vulnerability.]

Davis, M. (2001) *Late Victorian Holocausts: El Niño famines and the making of the third world*. Verso, London. [Stimulating and controversial; material on colonial attitudes to famine and natural disaster and on the role environmental hazards may have played in development. Argues that colonial intervention made peoples more vulnerable to climatic events.]

Gribbin, J.R. (1988) *The Hole in the Sky: man's threat to the ozone layer*. Corgi Books, London. [Good introduction to ozone issues.]

Harvey, L.D.D. (2000) *Global Warming: the hard science*. Prentice-Hall, New York (NY). [Review of global warming.]

Houghton, J. (1994) *Global Warming: the complete briefing*. Lion Publishing, London. [Introductory review of global warming.]

Leggett, J.K. (2000) *The Carbon War: global warming and the end of the oil era* (1st edn. published 1999 by Allen Lane, London) Penguin, London. [Global warming and fossil fuel use – political aspects.]

Whittow, J. (1980) *Disasters: the anatomy of environmental hazards*. Pelican Books, London. [Clear and comprehensive, if slightly dated, review of natural hazards with excellent illustrations.]

Websites

International Strategy for Disaster Reduction: http://www.unisdr.org/

Disaster and risk – political interpretations: http://www.anglia.ac.uk/geography/radix

Society for Risk Analysis: http://www.sra.org/glossary.htm

Network for Environmental Risk Assessment and Management: http://
www.eng.uwaterloo.ca/

(Accessed January 2004)

Part III
Environmental tools and policies

9 Environmental management methods, tools and techniques

Key chapter points

- This chapter reviews the methods, tools and techniques used by environmental management. Many of these are shared with other fields like environmental appraisal, impact assessment, quality assurance and many others. The diversity makes it difficult to give more than a brief overview.
- Each situation faced by an environmental manager demands the selection of a toolkit, which probably will need fine-tuning. With new challenges there may be no tried toolkit and it is then necessary to assemble one. The outlook and affiliation of the environmental manager will colour all this, although there may be some influence exerted by published or legally enforced guidelines, laws, treaties and professional bodies.

The terms 'methods and techniques' are frequently used but seldom clearly defined; I suggest that a method is a general manner of approaching something, and a technique is a specific application of a tool or approach. Methodology is like a 'battle plan' and techniques the 'type of weapon' used. A tool is something used to collect, analyse or present data or achieve some outcome.

Before the 1970s, few addressed environmental problems until after they had become manifest and problematic. Today, although it is still common to take no anticipatory measures, the trend is toward a proactive approach. So the environmental manager does not wait for problems or opportunities to manifest themselves; whenever possible efforts are made to look ahead. Consequently, predictive tools are needed.

Environmental management decisions should ideally be based on data acquired from more than one source, using a number of methods, and decision making should also rely on more than one method. The reality is that often poor data is available and there is little time or money to make rational decisions. Environmental managers can choose from a huge range of tools to support research, administrative activities, enforcement of environmental standards, communication with decision makers or the public and so on. Environmental managers have developed some of these, but many are borrowed from other disciplines, governance or business. The tendency is for practitioners of environmental management to specialise in particular activities.

Each situation or goal demands a specific set of tools, but as far as possible there should be standardisation so that successive results can be compared, and lessons learnt are reasonably transferable (Welford, 1996a; Thompson, 2002: 5). It is also useful to have an idea well in advance of need as to what sorts of 'toolkit' will suit particular tasks. Some of the more common categories of environmental management tools are:

- Scoping: setting limits to the task, developing terms of reference, identifying resources required;
- Strategic planning: setting goals and objectives;
- Identifying and organising the people with skills and knowledge, and finding the funds to conduct environmental management;
- Assessing risks and priorities;
- Assessing the options available for reaching identified goals;
- Implementing a selected set of options;
- Monitoring;
- Environmental auditing;
- Environmental guidelines;
- Environmental planning;
- Environmental policy making;
- Environmental enforcement/law: policing, legislation, taxation, penalties;
- Environmental research;
- Funding of environmental management;
- Life cycle assessment;
- Environmental performance assessment;
- Environmental reporting.

That is not an exhaustive list, and there is some overlap between categories. Environmental managers often specialise within one of a few of those categories. It is not possible in an introductory text to give in-depth coverage or critique of all environmental management tools; the following is a brief overview.

Types of tool

Abbreviations in square parentheses broadly indicate function as follows:

- Strategic management and planning [St]
- Data collection [D]
- Scoping and strategy selection [S]
- Auditing and monitoring [AM]
- Enforcement [E].

Green economics [AM] [D]

There have been efforts to make economics more environmentally sensitive for more than 30 years; the theory and practice of 'green economics' are still evolving. Given that the world is run by economic forces, and that companies are already very powerful and some look set to become stronger in the future than many countries, it is crucial that environmental management understands and works with economics. Through green economics, development might be steered, environmental management funded and accounting methods can be developed to track progress. There is now a discrete and rapidly growing field of green/environmental economics, overlapping with agricultural economics, natural resource economics, social economics and the greening of taxation (Costanza, 1991; Pearce et al., 1991, 1993).

Risk assessment and hazard assessment [AM] [D]

These tools are important and have long been used by the insurance, investment, war-gaming and gambling industries, and increasingly those interested in the impact of technology innovation and policy change. These tools are valuable for assessing threats in advance. The extent and frequency and type of hazard can be mapped; for example, a map of landslide or avalanche vulnerability. A risk assessment aims to identify a hazard and then quantify its significance and likelihood of occurrence: What? Where? How serious? How likely? How reliable is the assessment?

Once a risk has been assessed it remains to be established whether it is deemed 'acceptable'. When a country is desperate for electricity to heat homes in a harsh winter the risk of using a nuclear power station may be acceptable. Acceptable risk depends on people's perception, which is fickle; different ethnic groups, age sets, sexes and so on all vary in what they fear and will pay to avoid, but often opportunity to make a profit reduces fears. The insurance industry assesses risks and then, if it is profitable, finds ways for customers to spread the risk and impacts if they materialise wide enough to be acceptable, and give the company a profit.

Environmental managers are likely to face the following threat situations:

1   unexpected [U] – sudden
2   predicted [P] – sudden
3   unexpected [U] – gradual onset
4   predicted [P] – gradual onset

The worst scenario is the first; the second is quite a challenge, the third is more manageable *if* it is noticed early enough, and the fourth is the least problematic as long as it is perceived and action is taken. It is important to note that tools like risk assessment, hazard identification and environmental impact assessment are not precise and infallible, and that developers may ignore them or act on their advice in an inappropriate way. Care is needed to ensure tools generate no false sense of security or unwanted impacts (e.g. efforts to assess risk might cause panic).

There is growing usage of risk assessment and risk management by environmental management, in part linked to the application of the precautionary principle (Douglas and Wildavsky, 1982; Lave, 1987; Covello and Merkhofer, 1993; Norton et al., 1996; for further information see specialist journals such as *Human and Ecological Risk Assessment*).

Setting goals and objectives [S] [St]

There are various strategic planning tools, including brainstorming; policy research techniques; use of focus groups and social surveys to assess public or special interest group feelings; the Delphi technique, which orchestrates a team of experts, using controlled feedback, to get a pooled view on future developments; or use of specialists to research issues. The *Journal of Environmental Management and Policy Management* covers this field (available online at http://www.worldscientific.com/journals/jeapm/jeapm.html). Brainstorming may be via workshops, structured group meetings of stakeholders involved in environmental issues or focus groups; this differs from the Delphi technique, where the moderators are more passive. Environmental management frequently must use qualitative information from less than ideal sources. So it is important that environmental managers are aware of the limitations of their sources.

Scenario development [P] [St]

Environmental managers need to be proactive, which demands assessment of future scenarios. Scenarios are hypothetical sequences of events, constructed for the purpose of focusing attention on causal processes, crucial developments and for providing insight into ongoing situations. They are not accurate forecasts, hindcasts or backcasts, but rather projections reflecting different perceptions, which enable exploration of different responses. Scenarios can be constructed for the future or may start from a known state, tracing the route that seems likely to have led to it. Often scenarios are derived from modelling or brainstorming by a group of experts.

Ecolabelling [U] [E]

Ecolabelling is increasingly used on products to indicate how much they impact on the environment. The consumer can judge one product against another and, hopefully, buy the greenest. Various independent assessors undertake the labelling, with the focus on product impact. It therefore says little about manufacturing or recycling impacts. Policing and standardisation needs improvement.

Life cycle assessment [AM]

Often environmental managers deal with processes – manufacturing, running power stations, building and managing large projects like dams, and so on. It is very useful to know what the impacts and demands are throughout the life of something; impact assessment gives only a 'snapshot' view. Life cycle assessment, or life cycle analysis (LCA), first appeared in developed countries in the 1960s, and is a tool which seeks to identify impacts and demands at each stage of manufacturing, service provision and so on. Impacts do not cease when, for instance, goods leave a factory – there may be pollution associated with their use and disposal – and LCA assesses impacts for the whole life cycle. The public became aware of LCA in debates about disposal of product packaging, especially since the passing of legislation to require manufacturers to adopt LCA in Europe and the USA.

By the 1990s, the usage of LCA had diversified to include general support for environmental management; for example, by helping to identify stages in manufacturing or service provision where environmental measures are needed and are most effective. It is also a tool for helping environmental managers understand environmental problems. LCA can also be used to help evaluate the impacts and best practice at each stage in the provision of services, or in manufacturing or consumption – from raw materials to end-of-life disposal or recycling of products and decommissioning of a factory or other facility. In one study LCA showed that recycling of used packaging at the end point of cardboard production was less environmentally friendly than composting or incineration, because the re-use would consume much more energy than virgin fibre production (Stuart and Evans, 2002). Currently the UK is about to decommission nuclear power stations – little thought was given to the challenges this would offer when they were built in the 1950s and 1960s, yet much could have been engineered-in to help if the life cycle had been considered.

Some organisations have considerable experience with LCA practices (e.g. ISO 14040–14048 EMS – Environmental Management System – standards). Heiskanen (2002) noted that the spread of LCA was encouraging practices like design-for-environment, environmental labelling and others that seek to integrate manufacturing and service provision with environmental concern. By 2002 roughly half of large companies in Europe and the USA appear to have conducted some form of LCA (Heiskanen, 2002), so wider future adoption in developing countries looks likely.

Geographical information systems [D] [St]

Environmental managers need means to acquire, store, update, retrieve and display information. In the past maps, clear overlays and card indexes have been used; nowadays cheaper and more powerful computing has developed powerful geographical information systems (GIS) to gather, store and manipulate data. GIS can compare and display a huge range of data and it can be regularly updated, sometimes in real time, and may acquire data from existing databanks and monitoring systems. Once established, the GIS can quickly print out various data. It is easy to use GIS to seek correlations between various data sets.

Where developing countries have a limited database a GIS may be able to provide an alternative relatively fast and cheaply. Fed with remote sensing data, it can rapidly give a literally 'birds-eye view' and can monitor even remote and hostile areas. Most developing countries have established GIS units, and some have also developed their own satellite remote sensing systems, including China, India, Brazil and Nigeria; satellite remote sensing data can also be purchased from commercial sources or developed countries, e.g. Système pour l'observation de la terre (SPOT) or the National Aeronautical Space Administration (NASA), and from aircraft and remote instruments. GIS is a powerful and flexible means for predictive modelling, for planning, for checking compliance with regulations, and a research tool.

Remote sensing [AM] [D]

Information can be gathered from above the Earth's surface by satellites, aircraft, balloons and even kites. This may be in the visible spectrum or can be infrared, ultra-violet or radar imagery. Aerial imagery enables monitoring of large and remote areas, allowing national and international agencies to get a strategic view – much of the development of remote sensing was by the intelligence agencies of developed countries. Satellites and aircraft have registered smoke plumes from illicit forest burning, enabling prosecutions; oil slicks may be sighted before they spread too widely; wildlife movements may be tracked. Cheap unmanned 'drone' aircraft and simple powered hang-gliders can be used as airborne platforms.

There is also terrestrial or marine remote sensing: either static monitoring equipment linked to radio or telephone links, or sensors and transmitters small enough to be housed in buoys, submersible vehicles or even carried on or in wildlife. Such monitoring can provide constant data from a mountaintop or remote rainforest, or could allow monitoring of wildlife movements, body functions and data about their surroundings.

Participatory and rapid rural appraisal [AM] [D]

During the last few years there has been the development of a number of tools for monitoring and evaluating livelihood strategies, social parameters – e.g. social capital conditions, ability to innovate, attitudes, useful traditional knowledge and so on – and vulnerability. Aid agencies and non-governmental organisations (NGOs) have mainly developed these tools, with inputs from impact assessment specialists, anthropologists and sociologists. Monitoring and evaluation techniques used by the social sciences are valuable for environmental management because they show people's needs, vulnerability and strengths, and can uncover useful knowledge (Save the Children, 1995; Barrow, 2000).

Often results are wanted in a hurry and cheaply, so that even tools that are relatively 'quick and dirty' – i.e. fast but not very accurate or detailed – are valued. As a result, some techniques and tools seek to be rapid – like rapid rural appraisal (RRA). It has been fashionable for a couple of decades to assume that local people should be involved in data gathering, planning and decision making, so a participatory approach is now common. In the past a failure to consult people commonly led to negative social impacts which could have been avoided, and valuable local expertise was missed.

RRA is a methodology for rural development studies, developed since the early 1980s, which relies on researchers making in-depth and informal contact with people, observing local conditions and collecting other available data (Carruthers and Chambers, 1981). It is suited to investigating the numerous complex linkages involved in livelihoods. It places stress on the relevance, comprehensiveness, multidisciplinarity, speed and low-cost of data collection. It is much faster than most normal academic anthropological research (see the Rural Development Institute, online at http://www.rdiland.org/RESEARCH/Research_RapidRural.html, United Nations University at http://www.unu.edu/UNUpress/food2/UIN08E/ and http://www.developmentinpractice.org/abstracts/vol07v7n3a10.htm).

These techniques are also often multidisciplinary or even holistic. The goal is to get an idea of the needs, desires, capabilities and wishes of a group of people and to assess their socio-economic and physical environment (Shepherd, 1998). The approaches yield mainly qualitative and not necessarily accurate data, but also hold out the possibility of avoiding misunderstandings or promoting development that simply cannot be sustained or which is even unwanted. Gender analysis is an important component, vital because many environmental management issues are strongly gender-related.

Environmental and environmental management accounting [AM] [St]

Environmental (green) accounting is vital for environmental management and policy making. It has been defined as the gathering, ordering and presentation of environmental data expressed in economic terms. It is conducted at various levels – some countries have produced national environmental accounts, and it has been applied to various international issues, business and other organisations. There is overlap with environmental management accounting (Thompson, 2002: 246–64; Asia-Pacific Center for Environmental Accountability, available online at http://www.accg.mq.ecu.au/apcea; IUCN Green Accounting Institute at http://www.iucn.org/places/usa/literature.html, accessed January 2004).

The idea of eco-efficiency has appealed to some businesses – this has been defined as the delivery of competitively priced goods and/or services that satisfy needs and bring quality of life, while progressively reducing ecological impacts. This might be shortened to 'doing more, better, with less', and it falls well short of sustainable development because it does not address poverty and is too limited. Nor does it question the need for making a product or providing a service.

Environmental values and costs, capital and rate of drawdown of resources must be established, regularly updated and assessed, and clearly presented to decision-makers and ideally the public. Environmental management accounting is used in several different contexts, including management accounting, financial accounting and national accounting. It has been defined as the identification, collection, estimation, analysis, internal reporting and use of materials and energy flow information, environmental cost information and other cost information for both conventional and environmental decision making within an organisation. It is seen in some quarters as a useful tool for seeking sustainable development (International Environmental Management Accounting Research and Information Center, available online at http://www.emmawebsite.org/about_ema.htm, accessed February 2004).

Cost-benefit analysis [AM]

Cost-benefit analysis (CBA) predates most environmental management tools, dating back to the 1930s, possibly even the 1860s. It is allied to plans, projects, programmes and policies to try and calculate positive and negative impacts, in some cases in advance of a proposed development. It seeks to value impacts in economic terms, which can mean problems assessing environmental and social items – efforts to do so are usually indirect, using techniques like opportunity costs, shadow pricing and property values. CBA is a tool that is intended to help developers select from a set of defined development alternatives. It can be applied to projects, plans, programmes and policies. The results are given in monetary units, if need be using techniques like contingency valuation to try and estimate the worth of things that are difficult to value directly. One of the oldest tools – it has been in use since the 1930s – is well established and has generated a huge literature. In reality it is not as objective as many hope, and there can be difficulties valuing some things. There are ongoing efforts to update and modify CBA to improve its performance, none of which so far has cured all its faults.

CBA is less useful in developing countries because people there are more likely to operate outside of any formal market setting. Also, poor people may largely exist 'outside' economics – they consume what they produce and value social or cultural things, which have no clear monetary value.

Cost-effectiveness analysis [AM]

Cost-effectiveness analysis seeks to select development alternatives on the basis of lowest monetary costs – i.e. best value for money. A goal is set, say an improved environmental standard, and assessors seek the least costly way to achieve it.

Environmental impact assessment [AM]

Environmental impact assessment (EIA) and social impact assessment (SIA) are used to identify and evaluate the likely consequences of a proposed development, project, programme or policy. EIA and SIA – referred to together as impact assessments – adopt a structured approach, which should help prevent issues being overlooked. They have the potential to reduce negative environmental and social impacts and ensure opportunities are not overlooked (Sánchez and Hackling, 2002; International Association for Impact Assessment, available online at http://www.iaia.org and http://www.ext.nodak.edu/IAIA, accessed March 2004).

EIA has received more investment than SIA and is more widely practised. Both tools make use of a wide variety of techniques to make their assessments, including some discussed in this chapter. EIA and SIA can also be used to support sustainable development. They can cause developers to 'look before they leap' and ensure more accountability for their actions. Impact assessments can also help empower people, alerting them to proposals, and providing a means for them to participate in development decision making – at least this has started in richer nations. It is part of the move in Western, and increasingly other, countries away from top–down decisions by a remote technocracy who make little effort to consult the people, to a situation where citizens are made aware of options and consequences and given a chance to influence policy making and implementation. It is thus potentially much more than a tool.

Technology assessment or technology impact assessment is a subfield of impact assessment, which seeks to determine in advance of innovations what the impact will be (Porter, 1995).

Despite the increasingly global usage of EIA, SIA and environmental management systems (EMSs) (see later this chapter), and the common ground shared between them, their interrelationship is often poorly defined and should be strengthened. Impact assessment has the potential to reveal what the consequences of a development will be to developers and authorities, and possibly also to a public audience; EMSs, on the other hand, focus on what is required to implement adequate environmental management, to confirm its effectiveness and to ensure that improvements are made where it is possible and necessary, largely for an internal, management and employee audience (Sánchez and Hackling, 2002: 26).

In practice, impact assessments are often side-stepped, manipulated, or poorly heeded – improving links with EMSs might help counteract that. EIA is more likely to involve natural scientists and SIA social scientists, while EMSs tend to be dominated by business and industrial specialists. Both EMSs and EIA/SIA seek to reduce the negative impacts of human development, and each benefits from and assists the other – the former managing things to avoid problems, the latter identifying potential problems.

EIA and SIA, if satisfactorily conducted, should ensure:

- A reduction of unexpected impacts;
- More advance warning of problems;
- A systematic approach to development, which means all possible options are considered against possible impacts;
- Some degree of public involvement as a consequence of the release of impact statements;
- Increased likelihood that developers will be made aware of local environmental and social factors that might otherwise have been overlooked;
- More accountable, and hopefully more cautious, development;
- Awareness of social issues which may disrupt technical innovation or economic changes;
- Access to local knowledge that might otherwise be overlooked.

In short, impact assessment should do much to ensure more effective environmental management. Some claim impact assessment could actually help integrate various aspects of environmental management.

Unfortunately, with 30 years or so of hindsight, a significant number of critics argue that impact assessment largely fails to produce the aforementioned benefits for a variety of reasons, including faulty techniques, poor practice, weak enforcement of assessment findings and insufficient integration with environmental management. During the last 10 years a number of impact assessment specialists and bodies, notably the International Association for Impact Assessment, have worked to improve practice and enforcement, and there have been studies exploring how there can be better integration with and support for environmental management (Nitz and Holland, 2000).

One weakness of EIA and SIA is associated with identifying indirect and cumulative impacts. Indirect impacts are those lying along a chain of causation remote from relatively easily visible direct impacts; sometimes a number of indirect impacts combine to cause a cumulative impact. EIA and SIA are rather like cars on dipped-beam headlights: they can see the direct impacts but not much beyond, down the chain of causation. For some time there have been efforts to develop better indirect and cumulative effects assessment; one potential route is to make use of strategic environmental and social assessment and 'context scoping' (Baxter et al., 2001). Terminology is a little confusing, with cumulative impact assessment often used to describe tools which seek to predict cumulative impacts, and also there is cumulative environmental assessment, which likewise seeks to assess the cumulative effects of a proposed development. Although these tools are still being developed, a number of developing countries and some international agencies now use them before making decisions on development proposals (Munn, 2002).

Strategic environmental assessment [St] [AM]

Since the 1980s, strategic environmental assessment (SEA) has attracted the attention of planners, policy makers and managers. It appeared in the late 1980s as a development of EIA, which went beyond the local-focus, 'snap-shot' temporal-view and essentially project focus to consider the likely interaction of multiple projects, or the impacts of proposed programmes and policies, at the regional, national and even global level – hence the word 'strategic'. Initially this interested those involved in land use planning and regional planning, but it has spread to transport and aid planning, and to a growing range of policy-making and management fields, including environmental management (Lee and Walsh, 1992; Wood and Djeddour, 1992; Thérivel and Partidário, 1996; Partidário and Clark, 2000; Goodland, 1997; Noble, 2000; Verheem and Tonk, 2000; Fischer and Seaton, 2002; Jiliberto, 2002).

Impact assessors are trying to develop SEA as a way of integrating assessments to avoid a snapshot view, avoid duplicated efforts, and provide a means for checking the off-site impacts of one development on others. In practice, SEA is complex and challenging, involving assessment across multiple sectors at different tiers (levels) of assessment. There are hopes it will make the management of large programmes more effective and it has been promoted as an instrument for supporting decision making for sustainable development (Lee and Walsh, 1992). Some aid agencies have used SEA as a way of assessing the impacts of proposed modifications to their programmes, and it has also been applied to transport policies.

So far there is a lack of a clear SEA theory and a proliferation of different approaches, which has hindered its development. While there is still a lot of debate about the form SEA should take, and how it should serve to aid decision making and the quest for sustainable development, it does have promise. Europe issued an 'SEA

Directive' a few years ago (European Commission, 2000) which now requires plans and programmes to implement SEA measures. Developing countries are moving to establish SEA procedures (Caratti, et al., 2004). Broadly, there are two schools of thought in current SEA: one favours it dealing with the environmental aspects in the quest for sustainable development; the other seeks to incorporate social and other issues in addition to environmental issues. It is possible that SEA will become a useful way of integrating environmental management into wider policy making and planning.

Brainstorming [S] [St]

As soon as any development is mooted, it is usual for those involved to get together and actively discuss the approach, problems and opportunities, among others. One may call this 'brainstorming'. Using the Delphi technique provides a more structured brainstorming; this is a procedure for assembling and interviewing a group of experts who are given feedback without this being personalised, so that it further focuses discussion. It is essentially a way to get a group consensus and has been applied to gambling issues, war gaming and future scenario assessment (Barrow, 1997: 18). The technique is useful when a development is not clearly defined, and/or data are not very adequate. The results are not very accurate, but it is swift and relatively cheap; and has become easier with teleconferencing and e-mail.

Pilot studies [D] [St]

Pilot studies are a common-sense, simple approach to assessing a proposed development – a small-scale, possibly simplified forerunner to the project itself. They could be more widely adopted; however, they demand time and may thus delay the planning and implementation process. Also, results at a small scale may not necessarily scale up to predict the larger-scale situation.

SWOT analysis [St] [S]

This is a simple tabulation and *assessment* of strengths (S), weaknesses (W), opportunities (O) and threats (T), which offers environmental managers a useful overview of a situation or proposed development. It is relatively quick and simple and can feed into brainstorming.

Trend analysis [D] [AM]

Trend analysis is widely used and is a relatively objective and analytical approach, provided data are of reasonable quality. It consists of presenting and interpreting time-series data to show past patterns, which might allow future projections. The trends can also be useful for assessing performance and to derive indicators to benchmark with.

Benchmarking [AM] [St] [E]

Benchmarks are established levels which environmental managers, planners and administrators can use to judge standards, progress and compliance against. They may be internationally agreed units or standards or published descriptors. Benchmarking facilitates comparisons of performance so that there can be judgement whether a given performance matches what has been achieved before or elsewhere. It can be applied to virtually any activity, provided benchmarks have been prepared and agreed.

Eco-audit [AM]

Eco-audits (corporate environmental auditing) were first developed in North America in the 1970s to evaluate compliance of facilities and operations with laws, regulations and good practice. Eco-auditing seeks to establish if a company, government department, service provider, region or whatever is environmentally sound, and whether it is working to maintain and improve its performance. The end product is an audit report and an undertaking to regularly review progress and seek improvements (Barrow, 1999b: 64–74). Eco-auditing is a systematic multidisciplinary methodology, which seeks to periodically make an objective assessment of an organisation's environmental performance. Impact assessment – EIA and SIA – deals with potential impacts, eco-audit focuses on actual effects. It is already in quite widespread use in developing countries. An independent body usually accredits the process, typically a standards organisation like the International Standards Organisation (ISO), which can withdraw approval if standards slip.

Eco-auditing can offer the following:

- Provide data for environmental managers;
- Promote ongoing improvement by discouraging developers from damaging the environment;
- Help show if sustainable development is being achieved;
- Provide a means of regularly monitoring activities;
- Inform the public about activities;
- Identify opportunities for recycling, energy saving, use of by-products and so on;
- Help enforce environmental regulations;
- Provide information before a body is remodelled, sold, merged, etc;
- Check for wastage;
- Prompt proactive and contingency planning;
- Identify health and safety issues for checking;
- Possibly reduce insurance costs and legal liabilities;
- Improve staff morale and 'green image' of the commissioning organisation.

The majority of eco-auditing has so far been voluntary. Some countries are insisting on it and a few offer some funding assistance for smaller organisations and companies. Some NGOs have aided institutions to conduct audits, and major funding bodies may require it and often pay for it. India has modified its Companies Act to include an eco-audit requirement, and since 1995 Indonesia has required companies to undertake eco-audits. Business is also driving it – some large companies insist their suppliers and subsidiaries conduct it. It can help planners to develop contingency plans and reassure the public. Costs vary, depending on the complexity of the audit, location and whether it is a new development or similar to previous. Poorer institutions need funding to conduct audits (Barrow, 1997: 44–8). Eco-audit supports a proactive approach to environmental management.

Environmental management systems (EMS) [AM] [St]

Without reliable, widely accepted and understood standards, activities like eco-auditing would be ad hoc and suspect. As a result, various bodies have developed and promoted EMSs, the first being the British Standards BS 7750 which evolved from a total quality management (TQM – see below) standard – BS 5750 – and was first released in 1992 (see http://www.bsi.gobal.com, accessed January 2003). For example, the European Union Eco-Management and Audit Scheme (EMAS) is a site-specific and proactive approach promoted since 1995 (Barrow, 1997: 48; http://europa.eu.int/comm/environment/emas/ and http://www.eli.org/isopilots.htm, accessed June 2003).

Total quality management (TQM) originated in Japan, and has evolved into a business management field concerned with promoting organisation-wide quality – EMSs grew out of this. Organisations that embrace TQM tend to support integration of environmental management, health and safety management (Wilkinson and Dale, 1999), risk assessment and management standards.

Organisations can view environmental care as a cost and resist environmental management, or they could see standards and regulations as something to be met, in which case they tend to merely comply, or they can adopt an environmentally aware stance and show a genuine desire for improvement. A growing number of organisations run an EMS. An EMS can be defined as an organised approach to managing the environmental effects of an organisation's operations – it involves integrating environmental respect and awareness with economy and quality of production (Stuart, 2000). So, as well as achieving cost savings through environmental initiatives, an EMS allows an organisation to integrate environmental management into overall management (for an evaluation of EMS costs see Alberti et al., 2000).

An EMS enables an organisation to set goals and monitor performance against them, and shows when to take corrective action or make improvements if need be; also, it supports the development of a reflective outlook which seeks to be environmentally sound (Moxen and Strachan, 1995: 35; Hunt and Johnson, 1995: 5). Institutions and organisations select recognised standards with which to accredit their EMSs; currently the most widely used are the ISO 14000 series. The ISO 14000 series includes over 60 certification systems – ISO14001 to 14061 – which apply to eco-audit, life-cycle assessment and so on (see International Organisation for Standardisation standards relating to environmental management, available online at http://www.iso.ch/iso/en/stdsdevelopment/tc/tclist and http://www.iso14000-iso14001-environmental-management.com, accessed Feb. 2004; and http://www.iso.ch/iso/en/iso9000-14000/pdf/iso14000.pdf, accessed October 2003). Derived from the ISO 14000 series, and related to the ISO 9000 quality management series, some companies use ISO 9002 to certify their EMS. ISO 14001 was launched in 1996 and provides a framework and guidelines aiding the voluntary development of assessment and environmental practices. It also indicates what is needed for an EMS (scoping): the format, objectives, targets and implementation, review procedures, correction and so on. Unlike the European Union's EMAS or eco-audit accreditation, most EMSs make no requirement for a public environmental management statement.

Sometimes the EMS is conducted 'in-house', but it is most likely to be undertaken by the accrediting body or a subcontractor. The EMS process should be one of continuous, ongoing improvement, with a cycle of goals set, checks conducted and results published. Thus, the process should take a body beyond mere compliance and encourage it to become proactive and stimulate good practice. Use of an EMS should also help keep a body aware of changes in knowledge, legislation and so forth. An EMS should help ensure a structured, standardised and balanced approach to environmental management.

Supporters claim an EMS helps clients reduce environmental incidents and civil liabilities, increases the efficiency of operations by cutting waste and encouraging use of by-products, improves awareness of environmental impacts of operations, prompts assessments of hazards and threats and boosts corporate social responsibility. Adoption of an EMS should improve a body's 'green credentials' – image, attractiveness to employees and so on. Regulators are more likely to treat bodies using EMS with a 'soft touch', and management who use EMSs may enjoy greater peace of mind and pride.

ISO 14001 and its future derivatives will probably become the worldwide standard for environmental management (Schoffman and Tordini, 2000; Cascio, 2003; Osuagwu, 2002). In 1988 roughly 8,000 organisations used ISO 14001: 54 per cent were in Europe and many of the others in Japan – the USA has been slow to adopt EMS. ISO 14000 series EMSs are now emerging as the international standard likely to be increasingly used in developing countries. A random survey of European companies in 1998 showed about one-tenth had embraced EMSs (Baumast, 2001: 153). The ISO 14001 system is often adopted by organisations with little environmental management experience, in developed and developing countries.

There have been some spectacular cost savings claimed as a consequence of adopting EMSs: one ALCOA plant in the USA – a large aluminium producer – reported annual waste costs fell from US $8.33 billion to US $6.50 billion between 1995 and 1998 (Rondinelli and Vastag, 2000: 505). Another example is a division of Meridian Magnesium Inc., which reported a saving of about US $2.0 million soon after ISO 14001 certification. Increasingly businesses insist on suppliers or sub-contractors having EMS certification. Whether this could become a barrier to trade for smaller firms is so far unclear. Countries with tight social relationships like Japan seem to be more willing to adopt EMSs, while countries with more individualistic citizens, like the USA, may be less enthusiastic (pressures to conform seems to help prompt use of EMSs).

The growing adoption of ISO 14000 series EMSs standards means some degree of globalisation, uniformity of assessment criteria and less opportunity for the assessed to evade honest an objective assessment because an international assessor is involved. Eco-audit provides a 'snap-shot' view – the situation at a point in time and space. It is best incorporated into a structured EMS as a framework.

Current environmental management often fails to ensure legal compliance and allows 'window-dressing' – i.e. it puts in place a system or structure but there is not a requirement that environmental performance improves (Rondinelli and Vastag, 2000). EMSs are adopted but only certify each management system, not the level of actual environmental performance. There is also little point in an organisation or government developing an EMS if it has insufficient funds to address any problems revealed, which often happens in poor countries. Critics argue that EMSs may be a substitute for adequate environmental management and are often bureaucratic, mechanistic and insufficiently flexible. So EMSs like ISO 14001 might lead to mere compliance, not a will to improve. Also, the cost of adopting it could deter those with limited funds. For some, costs probably outweigh benefits; EMSs currently range from roughly US $10,000 to US $200,000 per site, plus annual ongoing documentation costs of US $5,000 to US $10,000 (Bansal and Bogner, 2002). Also, it is difficult to de-certify a body if it is granted EMS standard status and then becomes sloppy.

ISO 14001 is generally seen to be 'tried and tested' and is spreading, encouraged by funding bodies, insurance companies, shareholders, NGOs and others. One advantage is that it is a reasonably adaptable system, which could be good for developing country use because their conditions are very diverse and often changeable.

Tools for measuring sustainable development [AM] [St]

In 1992 *Agenda 21* called for the establishment of indicators of sustainable development. Some of the tools already discussed can be used to assess the extent to which sustainable development is being achieved – for example, eco-audits and SEAs. Sustainability indicators, if they highlight the real underlying causes of environmental damage, will deter wasted efforts treating symptoms or pursuing 'cosmetic cures'. Indicators may be used to show the current state or to flag pressures and problems before they fully develop, or be used as a 'symbolic means' of informing people.

Because there is no single accepted definition of sustainable development, and there are different strategies for pursuing it, it is difficult to develop a universally accepted measure of it (Victor, 1991; Pearce et al., 1993; Van den Bergh, 1996; Friend, 1996; Hanley et al., 1999; Briassoulis, 2001; Jasch, 2000; Velva et al., 2001; World Bank, 2003). For information on sustainable development indicators, visit: http://iisdl.iisd.ca/measure/compendium.htm (accessed September 2003) and http://www.sustainer.org/resources/index.html (accessed July 2003). Judging progress toward sustainable development demands prediction of the behaviour of complex socio-economic and physical systems, and not simply using extensions of established economic, social and environmental indicators. The likelihood is that a number of indicators will become established based on different understandings of what is most important. In general, composite indicators have replaced single-dimension indicators. Hanley et al. (1999: 59–62) critically review and compare a number of these, including:

- *Index of sustainable economic welfare*: this is a socio-political measure first proposed in 1989.
- *Net primary productivity*: derived from the ecological concept of carrying capacity, which is the maximum population of a given species that an area can support without reducing its ability to support the same species indefinitely.
- *Environmental space*: developed in 1992.
- *An extension of the Human Development Index*: there have been efforts to modify or develop this to complement the Human Development Index (HDI) (Sagar and Najam, 1998). The HDI was proposed by the United Nations (UN) Development Programme in 1990 and has become a widely used multidimensional measure of development. Since then it has been considerably modified and now makes some provision for assessing sustainability. However, the HDI has a long way to go before it measures sustainability and environmental issues adequately, and analysts like Neumayer suggest consideration should be given to developing a 'green index' to complement the HDI, rather than further greening the existing HDI.
- *Factor X concept*: this asks 'by what factor can/should the use of energy/ resources be reduced and still have the same utility?' (Robért, 2000: 251). This is a flexible way of monitoring and modelling how to extract more from the resources being used. It can be modified to ask 'By what factor must resource flows to affluent societies be reduced to allow the poorer societies to improve their living conditions?'
- *Composite index of intensity of environmental exploitation*: similar to the HDI (Desai, 1995).
- *Less-general, more-focused sustainability indices*: these have also been developed for specific ecosystems and sectors of activity; for example: a *sustainable land management index*, and a *sustainable agriculture index*. These have been prompted by doubts about the long-term viability of modern agriculture as consequence of the pollution by pesticides, herbicides, fertiliser runoff and heavy use of petrochemical energy inputs (Rigby et al., 2001: 465).
- *Indicators of Farm Level Sustainability* – might prove useful for highlighting the key inputs and practices, which hinder sustainability.
- *Ecological Footprint* (or ecofootprint) – measures how much land is required to supply a particular city, region, country, sector, activity or individual with all needs, usually expressed in hectares per capita. Effectively, it is a measure of 'load' imposed on the environment to sustain

consumption and dispose of waste. The concept is based on the idea that each individual uses a share of the productive capacity of the Earth's biosphere for both resources and disposal of wastes. It is essentially a measure of human aggregate ecological demand and it can only be temporarily exceeded or the productive and assimilative capacity of the biosphere is weakened (Wackernaegel and Yount, 2000). For example, the average North American had an ecological footprint in 1995 ca 4.5 times what it would have been in 1905 (http://www.sustainablemeasures.com/Indicators/ISEcologicalFootprint.html, accessed December 2003). The ecological footprint is thus a useful ecological accounting method for assessing the demands made by humans on various productive areas (Palmer, 1999; Wackernaegel and Rees, 2003; http://www.redefiningprogress.org; http://www.ire.ubc.ca/ecoresearch/; http://www.iclei@iclei.org – accessed August 2003 – give information on ecological footprints). It can help show where human demands are problematic, aid evaluations of what could be done to improve sustainability, and provide a framework for sustainable development planning. However, it has been suggested that it underestimates human impacts and shows only the minimum needs for sustainable development, without a healthy margin for error, which the precautionary principle seeks (Furguson, 1999).

Ecological footprinting can be pursued using a number of different models – for a company, island, region, sector of production, city or whatever. A valuable tool for use in the quest for sustainable development, it allows comparison of the impact of different components on the same aggregated scale. It also aids the assessment of eco-efficiency; i.e., whether an organisation is making the most of its resources and waste disposal opportunities.

Ecological 'footprinting' has been used by regional planners seeking sustainable development, e.g. for cities, regions or small islands (Barrett, 2001). It can also be used by businesses, cities and other bodies to measure their environmental performance (Barrett and Scott, 2001). Roth et al. (2000) explored the value of the ecological footprint approach for aquaculture and found some serious faults, including its 'two-dimensional' interpretation of complex ecological and economic systems. Other problems are its failure to recognise issues like consumer preference or property rights; that it offers a temporally limited 'snap-shot' view; that it may fail to allow for the fact that natural systems are seldom stable; consumption per capita, fashions, and settlement patterns change and this may not be registered; also, it does not help find the most appropriate path for human activities. However, it does seem to offer a graphic and easily communicated image; it is a relatively transparent tool; and it may stimulate creative thinking about environment and development issues. Ecological footprinting should be used with care, and in combination with other tools.

Integrated environmental assessment [AM] [St]

Integrated environmental assessment is an interdisciplinary process which seeks to collect, interpret and communicate the likely consequences of implementing a proposal (Van der Sluijs, 2002). It has been applied to global warming issues, acid deposition and other fields, and is similar to the Delphi technique, in that it draws on informed group opinion – but usually through focus groups (a group of informants who are interviewed in a relatively 'hands-off' manner, but asked set questions).

Industrial ecology [St] [E] [S]

To achieve sustainable development, society will have to become more aware of the need for, and be willing to support, environmental management. Industrial ecology may be one means for prompting such changes. It has been described as 'an operational approach to sustainability' by which humanity can deliberately and rationally maintain environmental quality with continued economic, cultural and technical development (Frosch, 1995; Erkman, 1997; Dunn and Steinemann, 1998).

Industrial ecology seeks to ensure industrial activity is not viewed in isolation from its surrounding environment, but rather is seen to be integrated with it (Socolow et al., 1994; Graedel and Allenby, 1995). It seeks to understand how the industrial system works and how it interacts with the biosphere, and then to use this knowledge to develop a systems approach aimed at making industrial activity compatible with healthy ecosystem function. The assumption is that industrial systems resemble ecosystems. Industrial ecology is probably easiest if practised in systems with clear boundaries: a geographical region; a specific process related to some material or energy, e.g. PVC production; a sectoral grouping of organisations; or a chain of management – such as the life cycle of product manufacture (see life cycle assessment) (Boons and Baas, 1997).

There has been interest in industrial ecology for at least 40 years but little achieved in practice, apart from in Japan. Since the early 1990s the approach has received renewed attention. It is something which could be actively promoted by business in both developed and developing countries, and there is a growing literature linking industrial ecology with environmental management, occupational health and safety, and planning (for further information see *Journal of Industrial Ecology*; Lowe et al., 2000).

Natural capitalism [S] [St] [AM]

This is a theory that grows out of the idea that the economy is shifting from an emphasis on human productivity to a radical increase in resource productivity. It is seen by some as a route to better resource use efficiency (http://www.natcap.org/sitepages/artS.php, accessed Jan 2004).

Total quality environmental management [AM] [St]

There are linkages between EMSs, on the one hand, and occupational and environmental health and safety management, quality management, life cycle assessment and industrial ecology on the other, prompting arguments for all to be integrated into total quality environmental management (TQEM) (Borri and Boccaletti, 1995; Winder, 1997; Weaver, 1999). TQEM has been applied to management of environmental processes, but has yet to spread widely.

Strategic environmental management [St] [S]

While many issues can be addressed at a local level, some sort of strategic overview is crucial: 'act locally, think globally'. A number of the tools and approaches already discussed seek to provide a route to strategic environmental management (Pentreath, 2000; see also *Strategic Environmental Management*, a journal published by Elsevier).

Ultimate environmental threshold assessment [AM]

Derived from threshold analysis, this seeks to identify and monitor developments to avoid exceeding some point beyond which serious, possibly irredeemable, changes occur. The thresholds may be local or global. National park managers originally developed the technique, which has potential for supporting precautionary planning (Barrow, 1997: 41–2).

Modelling [AM]

There is a huge diversity of modelling – computer models, physical simulations at small scale, and many other approaches. All seek to simplify without misreading natural processes, so that they can be better understood and possibly to make future predictions. Often a systems analysis approach is adopted. Modelling is constrained by the quality of the data fed in and the skills of the modellers; results need to be treated with caution, especially if modelling large and complex systems, as is the case with global circulation models.

Project planning and management tools [AM] [St]

Business, applied science and social studies and development agencies have developed a number of project-focused tools. These include logical framework analysis (LogFrame or LFA), borrowed from management studies and used to establish a structure to describe a project or programme and test the logic of the planning action in terms of means and ends. It focuses on how objectives will be achieved and what the implications of action will be. A diverse range of checklists and matrix construction techniques are used by impact assessors and managers to provide a structured view of a development (Barrow, 2000: 96–7).

Environmental managers need to establish limits, opportunities and threats. They have to determine the costs of proposals, best options and progress. And they must unravel what really causes problems. It is increasingly important to check future scenarios, which can often be assisted by reviewing evidence from the past. A huge palette of approaches and techniques are available – whether operating in a narrow sector, or with a broader remit, at local or global level – they must select what their experience and training and the evidence suggest is the right mix.

Agency guidelines [E]

Environmental management is often promoted by environment departments of NGOs and funding bodies like the World Bank, the Organisation for Economic Co-operation and Development (OECD), the Department for International Development (DFID) in the United Kingdom, the United States Agency for International Development (USAID), etc. For example, the OECD have provided guidelines for multinational corporations: *OECD Guidelines for Multinational Enterprises* (2000) available online at http://www.oecd.org.

Pursuit of sound data and judgement tempered by deadlines, funding and real world governance

Before roughly the 1980s, scientists tried to fully research issues before offering any advice. That goal is still held to by scientists and social scientists in theory. But in practice environmental management challenges, some of which may be unresolvable once fully manifest, often demand quick and inadequately researched decisions – some have called this the environmental manager's dilemma. To wait is too late, to act may mean insufficient knowledge. Commonly, funding bodies and governments or companies make a decision to act and then consult environmental advisers; the situation is one of damage limitation, rather than optimum development choice. This is also likely to lead to piecemeal, ad hoc development, when what is needed for sustainable development is a strategic overview (which can be enforced). In addition to local focus and short-term focus tools, environmental managers need strategic planning and policy-making tools and influence at that level. A few strategic planning tools are appearing and governments like the European Union and Australia are beginning to require strategic impact assessment.

Concluding points

- There is an ongoing shift to a more proactive approach, increasingly embracing the precautionary principle, and also seeking to make the polluter pay. This is affecting approaches and tool selection. There is also more emphasis on prediction and assessment of future scenarios.
- Environmental management tools are seldom precise and rarely infallible. Decisions should be based on use of more than one tool and care must be exercised to avoid developing a false sense of security.
- Computers are constantly improving and there has been the introduction of powerful tools like remote sensing, GIS and the Internet, which has made desk research easier. These innovations have helped environmental management.
- Tools have largely been invented in developed countries or by experts from those countries; they may need adapting; and may be irrelevant when confronted with

subsistence societies largely 'outside' economics. There is a fashion for participation and empowerment; this may work in a liberal Western democracy where access to information is relatively established, and governance is reasonably efficient and open, and the public are willing to be involved – in some developing countries it may not work under current conditions.

Further reading

Thompson, D. (ed.) (2002) *Tools for Environmental Management: a practical introduction and guide.* New Society Publishers, Gabriola Island, Canada. [Excellent introduction to tools – there are few other general books. But it is not available as a paperback.]

Newson, M. (ed.) (1992) *Managing the Human Impact on the Natural Environment: patterns and processes.* Belhaven, London. [Principles of environmental management covered from a physical standpoint. Introduces many of the basic tools and approaches. Not much focus on developing countries.]

Harrop, D.O. and Nixon, J.A. (1999) *Environmental Assessment in Practice.* Routledge, London. [Covers the identification, prediction and evaluation of potential biological, physical, social and health effects of development. Deals with a wide range of methods and techniques.]

Glasson, J., Therivel, R. and Chadwick, A. (1994) *Introduction to Environmental Impact Assessment.* University College London Press, London. [Overview of environmental impact assessment. Not focused on developing countries.]

Katz, M. and Thornton, D. (1996) *Environmental Management Tools on the Internet: accessing the world of environmental information.* CFC Press, Boca Raton (FL) and St Lucie Press, 100E Linton Blvd., Suit 403B, Delray Beach FL33483. [A source showing where environmental management information can be found on the Internet – USA focus.]

Websites

Environmental management tools – good bibliography: http://www.gdrc.org/uem/e-mgmt.html

Environmental management tools – UNEP site: http://www.unepie.org/pc/pc/tools/

Environmental management tools – European Environmental Agency: http://reports.eea.eu.int/GH-14-98-065-EN-C/en

Environmental management tools – International Chamber of Commerce, a world business organisation: http://www.iccwbo.org/home/state

Environmental management tools – EMAS, with focus on EMSs: http://www.inem.org/htdocs/inem_tools.html

(All accessed April 2004)

10 Environmental accounting, green economics and business

Technical innovation holds out the main hope of environmental improvement ... people will rarely change their behaviour or accept a reduction in living standards ...

(Cairncross, 1995: viii)

Key chapter points

- This chapter considers environmental accounting, the greening of economics, the role business plays in environmental management and sources of funding which can pay for environmental management.
- Since the early 1970s it has become increasingly clear that economic growth cannot continue in an unlimited fashion and must be transformed into something compatible with environmental care and sustainable development. This is something that has to be done quite fast.
- Economic know-how is as important to environmental management as ecological skills. Many single businesses command greater economic power than whole developing countries and some companies go beyond that.

Economic activity lies behind much of the world's environmental problems. In rural areas of some developing countries people still have 'subsistence economy' lifestyles, which involve little use of money and yet cause land degradation; here environmental managers still have to account for resource demands and various trade-offs between stakeholders. Efforts to manage the environment must be informed by economic issues. Economic tools are of as much value to environmental management as ecological tools. Also, environmental management efforts have to be paid for. It has become clear that environmental problems have accompanied development in liberal Western economies, centralised soviet economies, federal and non-federal developing countries, under democracies or dictatorships – the economic approach adopted by most governments have all tended to mismanage environment.

Apart from Thomas Malthus in the 1790s and early 1800s, there was little attention paid to environmental economics before the mid-twentieth century. From the early 1970s, interest in environment–economics issues has expanded, developing from various ideological standpoints. One of the more influential early efforts is that of Schumacher (1973), who adopted a 'Ghandian' small-scale focus. There are a diversity of proposals for the 'way forward': reassurances that 'the market' will act to counter environmentally damaging activities, calls for strict economic controls to protect nature, and arguments that voluntary regulation will be sufficient. In all probability a mix of measures and approaches will be needed.

Economic growth and the environment

Since the 1980s it has increasingly been accepted that economic growth does not have to be at the expense of the environment. In a world where money is usually the key to achieving things, the role of business in environmental management is important (GEMI, 1998). With growing globalisation, multinational and transnational companies commonly play a central role in activities which affect the environment, and many, perhaps half, of the world's richest institutions are businesses not governments, generating much more revenue than any developing country can hope to muster. Environmentalists often resent economic growth – this is, as Cairncross (1995) pointed out, inevitable – so it is important to explore how its negative impact can be minimised. Some companies have changed to become environmentally sensitive – Jansen and Vellema (2004) presented case studies showing how agribusiness has responded to environmental issues and food production demands and how it can be harnessed for better environmental management.

Since the publication of *The Limits to Growth* (Meadows et al., 1972), there has been growing support for pursuing broad 'development' rather than mainly a growth (economic expansion) focus; financial indicators are still important benchmarks, but few measure progress any more solely by gross domestic product or export figures alone. Assessors now seek to establish if there has been an improvement in quality of life and sustainability, which are closely associated with avoiding environmental problems.

Poor environmental management can damage economic growth: polluted, unhealthy people are less productive labourers and there is likely to be a plethora of other costs. Poor people may have no choice but to degrade the environment as they struggle to survive – poverty and environmental quality are interlinked. A large proportion of the population of developing countries is poor, and tends to suffer more from environmental degradation than richer people. Also, environmental taxes on a per capita basis are likely to hurt the poor more than developed country populations.

Research and development are to a considerable degree driven by commercial forces, perhaps more than ever before. This means that solutions to problems involving the poor may be underfunded because there are limited profits to be made; it also means non-applied studies get neglected – and in the past these have been crucial for environmental management: for example, identification of the ozone hole over Antarctica and monitoring which established atmospheric carbon dioxide trends were not funded as commercially promising research.

Bodies like the World Bank now have environmental departments (http:// www.worldbank.org/environment, accessed March 2004); however, there are still many businesses that strive to maximise profits to show economic growth and pay shareholders – many of the latter in developed countries. Only in the last few years has the idea of ethical investment – allowing investors to direct their money to green activities, or those acceptable to pacifists, supporters of Islam and so on – become practical. There are signs that shareholders may shoulder more of the costs of environmental damage and insist on greener activities. Shareholders have started to ask questions on environmental policies at company board meetings.

Welford (2000:12) observed:

> ... The drive for economic growth and hunger for Western levels of consumption in the newly industrialising countries and the ex-Communist world are developing precisely at a point at which consumerism in the West is beginning to appear socially self-defeating and ecologically unsustainable.

Environmental controls are still often seen by business and some politicians as 'green tape', slowing development and raising costs. But the market alone is unlikely to deliver a less-degraded environment, although business could play an increasingly green role (Barlow and Clarke, 2002).

Some of the commitment by business to environmental management in the last few decades has been rhetoric, and the question increasingly being asked by environmentalists is 'Will sustainable development be adequately pursued before the collapse of established market economics – which, many predict, is going to be caused by environmental degradation?' With the previous question in mind, it is wise to nurture business to seek economic growth *and* support environmental management and sustainable development. This will be especially important in developing countries, where there is pressure for economic growth to counter poverty, and limited tax revenue. Growing food for today often means leaving a desert for the future, and current employment is frequently based on resource exploitation. Some countries have clearly improved their infrastructure and welfare by depleting their natural capital, such as tropical forests, metal ores and so on. A shift to more sustainable strategies is needed; one route might be to prompt business to reinvest more in activities that are environmentally sound.

The argument is often made that environmental resources are in least danger of degradation when they are privately owned, rather than being common resources or in public ownership, which is often the case in developing countries. When environmental resources in private ownership become scarce their price can be assumed to rise so owners take better care of them and nurture them, and consumers are encouraged to seek alternatives – the ' invisible hand of the market' hopefully comes into play and counters overexploitation. For elements of the environment not in private ownership there is no regulation, although there may be traditional controls.

There has been concern voiced about common resources for over 30-years. Since the 'tragedy of the commons' arguments of the late 1960s (Hardin, 1968; http://www.dieoff.org/page95.htm, accessed 2003), environmentalists and natural resource planners have continued to ask whether collective management of common resources can work (Berkes, 1989; Orstrom, 1990). In Europe, common resources like land had started to shift to private ownership before the sixteenth century. In the United Kingdom, by the eighteenth century enclosure and privatisation of common land was a driving force for the economy and caused thousands of rural folk to relocate to urban areas; these were developments which helped support the Industrial Revolution. Similar 'enclosure' is currently underway in many developing countries; the state or senior administrators sell licences for land, minerals, logging, fisheries and suchlike. Traditional users usually have no legally enforceable claim, even though they may have a long history of use, so displacement and marginalisation often result (The Ecologist, 1993). The multinational corporations or large national businesses which acquire such licences provide profits for shareholders, foreign income for the developing country and cash for ruling elites, and exploit resources which the host nation would otherwise probably be unable to develop. Resource exploitation can also be used to strengthen national claims over territory, as a display of sovereignty and progress.

Sometimes it may be state authorities in a federal system that support exploitation against national government wishes. Often the officials involved stand to profit and do not have a long-term, environmentally friendly view. But there are cases where this sort of approach seems to have been beneficial: the Falkland Islands, acting on advice from London University advisers, charge fishing vessels in their waters a 28 per cent levy on their catch as a fee. Without such revenue the Islands would not be able to afford to police the fishing, conduct research, monitor stock or manage conservation; the revenue has also made the Falklands relatively affluent (Cairncross, 1995: 77–8).

Natural resources are often under no clear and enforceable single ownership or even national sovereignty. There have been recent claims that sea fishing is so poorly controlled that 'the final roundup' is currently taking place – i.e. overexploitation is unchecked. There are indications that nine of the world's 17 largest ocean fisheries were being overharvested in 2003 (New Scientist, 2003). World fish stocks are in a poor state and time for developing workable agreements and enforcement is scarce. Unilever, Europe's largest fish trader, recently established a Marine Stewardship Council to support an approved trader label that indicates that fishing companies catch their products in a reasonably sustainable way. It is interesting to see the initiative coming from the business sector; unfortunately, it is nowhere near enough, and what has appeared has been judged ineffective and open to abuse. Soon, it seems, popular staple fish like Atlantic cod will be on the CITES Endangered Species List (see http://www.cites.org/).

Currently there is a sort of 'enclosure' or privatisation of common resources, like genetic material via patent law and claims of ownership of intellectual property rights. This could accelerate and disadvantage developing countries. There have been many attempts by business to uncover and control traditional knowledge in common ownership using ethno-botanists and social scientists (Berkes, 1999). For example, there have been protests at attempts by companies to patent products clearly based on folk remedies associated with India's neem tree (*Azadirachta indica*); the raw material and ideas for its use are 'stolen', a product is synthesised and then sold at a profit, and attempts are made to protect the trade by patent. As biotechnology develops similar issues are likely to increase.

A number of resources are exploited without a reasonable charge being made – water is often treated as a virtually free resource and little income is generated from it to pay for its environmental management, perhaps because often it is not communally owned. Water charges are commonly inadequate to pay for supply, discourage wasteful and environmentally damaging overuse, or generate income for combating environmental problems. Often the crops grown by irrigation schemes fetch profits that are too low to make it feasible to pay higher water charges. This appears to have been the case in parts of Cyprus in the late 1990s, where either the economic forecasting was inaccurate or government was happy to subsidise water prices paid by commercial agriculture. In 1992 at Dublin an international declaration on water stressed that it should be treated as an economic good.

Perhaps the worst case of exploiting water at the cost of the environment has been the Aral Sea region, where central planners, advised by some of the world's best hydrologists and soil scientists, failed between the 1960s and late 1980s to realise that the sale of irrigated cotton nowhere near compensated for the loss of fisheries and the severe environmental degradation its production caused. Environmental scientists warned of creeping degradation but were not heeded.

It is not unusual for developing countries to subsidise the cost of agrochemicals and electricity, but this can discourage good soil husbandry, prompt greater pollution and act as a disincentive to more efficient energy use. As global trade agreements discourage subsidies for agriculture in some developed countries there is a risk that commercial growers will shift to developing countries with fewer restrictions and cheaper land and labour, and cause environmental damage. What is needed is an environmental management overview which warns of likely problems and spots opportunities and solutions.

Globalisation, modern market economies, and 'consumerism'

The world is increasingly globalised – there is little to be gained from seeking a precise and universal definition. The development of media which reach across national borders, especially the Internet and satellite TV, have helped to support this globalisation trend. Regional and worldwide co-operation and trade agreements are also reinforcing the process. Fashions now spread faster and more widely than they did in the past and there is growing global interdependence for markets, raw materials, component manufacture and labour. All countries, but especially developing countries, are becoming more dependent on others.

Globalisation is presently dominated by an essentially 'modern', Western, capitalist, neo-liberal, 'democratic' culture and linked economics. This is often associated with the spread of humanism, secularism, abandonment of traditional ways, neglect of community and loss of social capital. The driving force so far has mainly come from the USA, Europe and Japan; the likelihood in the coming century is that there will be a shift to it being driven from southern Asia and the Pacific Rim, especially China and nearby countries. Rapidly growing linkages and developments in trade and economics are often outpacing international agreements and laws that control environmental impacts. For example, arguments can be made that the General Agreement on Tariffs and Trade (GATT) has favoured free trade over environmental protection. There has also been friction because World Trade Organisation (WTO) agreements have been seen to outlaw measures designed to assist small farmers; the risk is that these producers will be forced into degrading the land to survive. Recently, the Windward Islands (Caribbean), which are very dependent on banana production by small growers, has been forced to drop assistance measures from 2005, because the USA argued the aid was an unfair advantage (Myers, 2004).

As well as trade agreements, the world's citizens are increasingly demanding material possessions; this 'consumerism' is being fuelled by advertising and by the media presenting 'lifestyle' images people aspire to. Many environmental activists are deeply concerned that globalisation and consumerism conspire to threaten any hope of sustainable development and the maintenance of adequate environmental quality. Welford (2000: 56) summed up the present situation succinctly:

> There is now ... a dominant corporate culture which believes that natural resources are there for the taking and that environmental and social problems will be resolved through growth, scientific advancement, technology transfers via private capital flows, free trade and the odd charitable hand-out.

Some environmentalists urge a robust response – to establish a 'postmodern' and green worldview/culture to replace the current globalisation-plus-consumerism-militarism. Many feel that satisfactory environmental management and sustainable development depend on a more proactive approach, not a reaction to problems embodied in the present support for a 'business-as-usual' eco-modernism approach – i.e. the assumption that human nature and economics will not change much in the future, so environmental management will have to work with that – and technical 'quick-fix' reactions. They also argue for a fundamental shift in outlook and ethics.

Gradually, virtually all countries have come to accept that they are part of one interdependent global environment. While the world is far from agreeing enough workable environmental controls and revenue-raising measures, these challenges have been attracting growing attention (Jacobs, 1991; Helm, 1991; Daly, 1992; Van Weizsacker and Jessinghaus, 1992). More go-ahead environmentalists argue that ethical and social change is crucial and must be got underway fast; Welford (2000: 22) observed '... we ought not to see sustainable development as an end and an aim but as a mode of action which, in turn, has to be addressed in different ways by organisations operating with differing objectives and cultures.'

Many developing countries and businesses see economic growth as the route to development; however, it has often by-passed large portions of populations, has frequently failed to improve infrastructure and services, has done little to improve law and order or access to human rights, and has failed to maintain environmental quality. The goal of progress has to be redrawn to emphasise environmental quality and improved human well-being.

Environmental accounting

In an ideal world, authorities would ask 'What are the most pressing environmental problems?' and then give priority to spending to avoid, mitigate or adapt. However, we will never have perfect forecasting because environmental knowledge is incomplete and environmental, social, political and economic interactions are so complex. Spotting priorities is not easy. Even if problems are identified it may be a challenge to prove causation, estimate costs of solutions and compare these with the likely expense of inaction. Many environmental problems involve more than one developing country, some involve a mix of developing and developed, and some are global. The world now has fairly uniform, agreed units – compared with the chaos of imperial, metric and indigenous units peoples used before the 1970s (if anyone at all bothered to measure environmental parameters). The less-good news is that nations are still a long way from agreeing the values of human life and health, international law has even further to go to adequately embrace environmental issues than national laws.

Environmental accounting should help environmental managers assemble information and take stock. Environmental accounts bring together economic and environmental information in a framework to measure the contribution of nature to the economy and the impact of the economy on the environment (Rietbergen-McCraken and Abaza, 2000; Lange et al., 2003a, 2003b; IUCN Green Accounting Initiative, available online at http://www.iucn.org/places/usa/literature.html, accessed February 2004). By the 1980s, the United Nations (UN) was promoting an internationally agreed framework for recording what was happening with countries' wealth, health, etc. However, in practice, even though there are now multidimensional indices available to measure environmental health and progress with sustainable development, the application in developing countries is limited and patchy, and it is not easy to measure and monitor natural capital.

Several countries have established environmental accounting procedures, so far mostly developed nations (Gray et al., 1996). Recently, new indices have started to appear, and should help those trying to evaluate environmental costs and benefits and assess performance. Currently, developing countries, aid agencies and funding bodies are reviewing, testing and adapting these (Perrings and Vincent, 2004; Bartelmus and Seifert, 2003; Lange et al., 2003a, 2003b).

Many of the attempts to develop environmental accounts have adopted a regional scale approach, using watersheds, river basins, islands, coastal zones and similar. Bioregional units can offer discrete and stable biogeophysical units which also include a good deal of human activity, and which make sense in other ways (Sale, 1985; Stolton and Dudley, 1999: 41–9). Such units also support an integrated approach to study, planning and goal identification, monitoring, and management and allow a focus on local issues. There are presently signs that regional environmental management systems, integrated regional environmental management, integrated river basin development and management and similar strategies which address territory and function are attracting renewed attention; for example in the Lower Mekong Basin. A few years ago some developing countries rejected these approaches because they seemed to be too authoritarian and dated – hangovers from the USA Tennessee Valley Authority of the 1930s, which had given mixed results.

Natural capital accounting is difficult but sustainable development demands it, and it must be effective and integrated with environmental management (Turner, 1993; Eckersley, 1996). There is a pressing need to develop and put in place effective methods to measure and monitor sustainable development and to do so in a way that can be understood by economists, commerce and administrators; the most promising approach is probably an economic–environmental accounting framework.

Where an environmental asset is not being brought or sold it is not easy to give it a value. Non-market environmental valuation is widely practised (Haab and McConnel, 2002). There have been efforts to focus accounting on those parts of the natural environment which perform important and irreplaceable functions – critical natural capital (Chiesura and de Groot, 2003). People tend to be unwilling to pay if they do not directly benefit. Yet sometimes there are situations where people accept a cost without much immediate personal benefit. For example, they may contribute to wildlife conservation and few will ever set foot in Amazonia or Kenya. It is possible for a population to spend large sums of money to support football, which is not crucial for human survival, but balk at taxes to conserve biodiversity, control soil erosion or pollution.

Valuation is far from precise. Often it has to rely on indirect indicators – like the value of property, which is likely to be greater in a 'nice' area (contingent valuation) – to get an idea of what people are prepared to pay. Or focus groups and questionnaires can be used to check what people value. Peoples' values can rapidly change with fashion, the general state of their national economy, and aesthetic and religious outlook. A state can attach high value to something that would otherwise be worthless if it is judged to be of heritage or strategic value. Environmental valuation is also likely to vary between countries at different stages of development, and in a given nation through history. It is also likely that in seeking agreement on transnational issues there will be different views and priorities. So environmental valuation is not absolute. Carson (2004) provides a bibliography and case studies of contingent valuation use in over 100 countries.

Understandably, developing countries are likely to accept more environmental damage if their people see it as a 'price to pay' for jobs or facilities like roads, schools or hospitals. Since the 1980s developing countries have faced declining commodity prices, disadvantageous terms of trade, often the burden of servicing debt and of structural adjustment programmes, and marked population growth; they are then asked to conserve resources and avoid environmental degradation (Mackenzie, 1993; Reed, 1992; 1996). Political as well as economic forces come into play, and are often difficult to separate. Differences in style and capacity of government also give rise to divergent valuation (Cairncross, 1995: 35). As mentioned in earlier sections, where the benefits of environmental protection, and the expenditure to achieve it, accrue slowly and benefit those yet to be born and those in a different country, people often question the costs. Attempts to apply economic valuation to environmental problems in developing countries have been examined by Georgiou et al. (1997). The potential value of biodiversity from developing countries was mentioned in the last chapter. Swanson and Barbier (1992: 3) and Polasky (2002) provide some estimated market values in the mid 1980s for things like wild perennial maize, which is of huge value to crop breeders and could be worth billions of dollars.

Green economics

Since the 1980s, there has been considerable progress in making economics more environmentally focused. Economics before the 1980s mainly approached environmental issues through shadow pricing and attempted to incorporate environmental values into cost-benefit analysis (CBA) (Lowe and Lewis, 1980). Often, environmental factors proved difficult to value, and were simply treated as intangibles and effectively ignored. Things began to change in the late 1980s with the publication of the first of a series of books on green economics (Pearce et al., 1989). The field has subsequently flourished and has embraced sustainable development (Pearce and Turner, 1990; Daly and Townsend, 1993; Pearce and Barbier, 2002). In 1991 a Business Charter for Sustainable Development agreed a code of conduct, which it was hoped would be widely respected; this is now managed by the World Business Council for Sustainable Development, and cannot be said to have been especially influential.

How much should a developing country be reasonably expected to pay for environmental management? In the mid 1990s (according to Cairncross, 1995: 59) most developed countries were spending 1 per cent to 2 per cent of their GDPs on environmental protection – and the expenditure was rising. Assistance from richer nations for environmental management through bilateral aid, funding from NGOs and so on is available – but not as much as many would like to see. Developing countries are likely to have to rely on international money from sources like the Global Environmental Facility, and these are not always as well funded as they should be, or to the level contributors have pledged.

A question raised a number of times in recent years is whether developing countries have a reasonable damages case against former colonial powers for environmental degradation which took place decades ago. Cases which have resulted in settlements or ongoing claims for legal damages include some Pacific islands, where phosphate mining removed topsoil, and actions brought by Aboriginal peoples in Australia and the Pacific who suffered resettlement during 1950s atomic weapons testing. There are parallels in current claims against German companies which used slave labour in the Second World War, and the descendants of slaves in the Americas, a few of whom have voiced an interest in seeking reparation from former slaving nations. The question of whether a present-day company or government can be held responsible for actions that took place long ago is still far from resolved; company officers and shareholders who benefited may have died long ago. In the USA, at least, polluters are deemed responsible for past actions, even if they did all they could at the time to prevent it. However, the precautionary principle is difficult to enforce where there is no clear negligence, and where no individual polluter(s) are clearly identifiable.

Unforeseen problems arise during developments, and can be difficult to resolve after completion. Insistence on pre-development environmental and often also social-impact assessment is now very widespread, although the approach is easily sidestepped or corrupted. A few countries have laws which require a developer to pay a sum to be held in trust as an indemnity, for five or more years after a project is completed, to resolve environmental and other problems. However, when negotiating with companies to secure a project, some governments shelve such legislation rather than risk losing the hoped-for fruits of development.

Environmental taxes

Taxation can provide revenue for environmental management and serve as a 'command approach', prompting and enforcing what a government seeks. In the past, the usual strategy was to tax companies or individuals for environmental misdeeds; nowadays there is more likely to be a 'partnership' between the state – usually an environmental agency – and the potential taxpayer, whereby the latter seeks to co-operate and by so doing reduce taxation. Whether rigidly enforced by authorities or conducted through some form of co-operation, taxation depends on the establishment of suitable environmental standards, adequate monitoring to reliably identify when standards are exceeded, and effective enforcement through some form of tax punishment or reward for desirable conduct (Wallart, 1999; Russel, 2003).

Some may see environmental taxes as 'paying for the right to pollute', and this may effectively be the case if they are set too low. Ideally, authorities should support forms of taxation which discourage environmental damage, rather than punishing bodies which commit it. The reality is that developing countries may find fair and effective taxation a challenge for a number of reasons: limited funding for monitoring and enforcement; a pressing need to provide employment, even if it means some environmental damage; limited infrastructure, which hinders inspections; vulnerability to pressures from powerful companies and financial interest groups; and the potential for corruption. Any country can be tempted to argue that development is necessary for strategic reasons – and use it to side-step controls like taxation.

Frequently the problem is to ensure that environmental taxation is used for environmental management, and is not diverted for other things. Poorer countries may fear that green taxes could harm their competitiveness. Also, most taxation to date is of the standards and charges type; there are few situations where tax rates are equivalent to the marginal cost of the environmental damage at the optimal level of damage – so-called Pigouvian criteria (see Glossary). Pigou, a Cambridge economist, explored ways of estimating damage to the environment caused by pollution, so that it could be taxed and discouraged, as early as the 1920s (Pigou, 1932). Pigouvian taxes are difficult to calculate, and optimal levels may be unknown because many environmental processes are not yet fully understood, or are too complex to model. Also, taxation can be difficult when many problems are transboundary in character, i.e. cross borders and involve more than one country.

There is growing experience in environmental taxation in developed countries, but less has been done to develop it in poorer countries, and for those with different legal and religious systems from the West. Most environmental taxation is just grafted on to existing regulations. Environmental controls and taxation need adaptation to suit non-Western situations (Dien, 2000; Faruqui et al., 2001), not least because people must understand, respect and support controls.

Each country has different population structures, energy consumption patterns and natural resource endowment, which makes it difficult to fairly agree measures like internationally applicable taxation. A few years ago, a dispute about apportioning blame for carbon dioxide emissions raged between the World Resources Institute (Washington), seen to be promoting a line favourable to developed countries, and the Centre for Science and Environment (New Delhi), arguing on behalf of developing nations (Agarwal and Narain, 1991). Poor countries may also come under pressure to support rich nations in a host of ways, one of which may be voting a particular way in a forum, like the UN General Assembly, in return for aid. Funds may also be 'tied' – that is, granted with conditions attached, usually a requirement for the recipient to use some of the loan or aid to purchase equipment from the donor, or to supply raw materials or offer some other benefit.

One approach which has spread since the mid 1990s, by which time it had been reasonably established in the USA, is tradable pollution quotas – marketable permits or tradable credits. These work by setting a level or standard and allocating a quota to the potential polluter, fishing fleet, etc. Any emissions or catch above that standard are penalised, but if the country or business does not use its whole quota it is free to sell its surplus, perhaps by auction, or possibly save it for a later date. An authority may need to regulate how tradable quotas are sold to prevent speculation and hoarding. The approach is useful for transboundary issues, and negotiations to control global carbon emissions and regulate ocean fisheries exploitation have explored tradable emissions quotas.

Other funds for environmental management

In poor countries the chances of obtaining tax revenue from the general population for environmental management are limited. Some funds might be generated by taxing the rich on luxury goods, levying fees on tourists, taxing manufacturing, and from lotteries, the export of resources or through exploitation licenses granted to multinational corporations. Revenue can also be generated from nature-based tourism (green tourism), especially ecotourism; the latter can be based on wilderness travel, trekking, wildlife observation, camping and so on. The Congo Basin funds over two-thirds of its estimated recurrent costs of conservation through ecotourism (Chambers et al., 1996; Stolton and Dudley, 1999; Wilkie and Carpenter, 1999). Money for environmental management can also be obtained from debt-for-environment swaps and debt-for-nature swaps, hosting carbon sink plantations, or specific eco-aid grants.

Market-based approaches to environmental management have great potential in a world where lack of public funding means initiatives must pay for themselves (Swingland, 2003). It is important to build partnerships between locals, NGOs, business, international agencies and so on. But caution is needed; for example, the 1997 Kyoto Protocol included agreements for rewarding those who sponsor reforestation of deforested areas – but this could lead to speculators clearing unspoilt land and then restoring it for a tidy profit if no safeguards are implemented.

There have been calls for developed countries to levy taxes on their citizens to pay for the conservation of global commons and possibly some developing country environmental management; Cairncross (1991: 71) estimated that a levy of US $8.00 per person in rich nations would adequately pay for Amazonian protection, and the region is of worldwide value. If it is handled in a skilful way with careful marketing the world's richer citizens would probably willingly pay much more for environmental management beyond their own borders than is presently the case; the key is to convince people that they ultimately benefit and have a moral duty. There may be other sources of environmental management funding, in the form of 'international levies'. For example, a charge could be made for communication satellites and distributed to developing countries for environmental care; all inter-country gold exchanges could be subject to a charge to be distributed by the UN.

Conservation initiatives often seek to recoup at least some of their costs; in addition to ecotourism, there could be the sale of surplus game animals like alligators, souvenir materials or safari fees for visitors. Perhaps the closest link between conservation and environmental protection and exploitation to fund such activities is the extractive reserve. There is also a controversial field of patent charges and royalties applied to biodiversity; commerce, developing countries, NGOs and international agencies are involved in ongoing lobbying and debate about this.

When dealing with environmental management it is important not to lose sight of the fact that viable strategies demand more than monetary funding; they also need natural capital (the environment and natural resources), social capital (institutions and practices which support individuals and families), cultural capital (pride in surroundings, ability to organise, and know- how, and often technology) (Welford, 1996b, 1997; Berkes and Folke, 1998).

The value of local knowledge for supporting biodiversity conservation has been discussed earlier; it is also invaluable for improving soil and water conservation and food production (Reij et al., 1996). Too often during the past, local knowledge has been overlooked, misinterpreted and undervalued, a problem that persists today. With many people too poor and too far from the beaten track to adopt high-tech measures which demand inputs, the only hope is for them to use simple, low-cost approaches which utilise locally available virtually free materials and build on indigenous knowledge. Alternative, intermediate and appropriate technology, participatory appraisal and studies of indigenous knowledge have come together since the 1980s and should help promote better approaches for poor people.

Green business

Applying standards and regulations to, and monitoring, thousands of households and millions of individuals is a challenge; it is easier to seek environmental goals through the medium and large companies that serve those millions (Cairncross, 1991: 95). On the other extreme are those who attach less value to commercial efficiency and seek post-industrial alternatives to corporate globalisation as a route to sustainable development (Milani, 2000). Businesses vary a great deal in their impact on the environment, the resources they have available for environmental management and their outlook.

More extreme environmentalists tend to write off all business as exploitative; however, some companies are run by people who do care for nature and there are some advantages in going 'green'. Many companies have huge resources, both financial and in terms of expertise and ability to lobby governments, well in excess of anything that can be mustered by developing countries. Companies involved with potentially damaging activities can no longer afford to risk legal action, bad publicity, disillusioned investors, refusal of cover by insurers or loss of government licenses – environmental management has become something they cannot ignore. In the past, business was often keen to oppose, side-step, pay lip-service to or reluctantly comply with environmental controls. There are still companies that see environmental management as a cost and a burden, as do some developing country administrators; some hide behind a false facade of green publicity, which has been termed 'corporate greenwashing' (Welford, 1997). Without effective and transparent environmental accounting, such greenwashing disinformation is easier. Improving media and Internet communications also help counteract greenwashing by making it easier for environmentalists to find and exchange information and attack offenders. Various NGOs discourage greenwashing by public ridicule; for example, making regular greenwash awards (see Corpwatch website at http://www.corpwatch.org/campaigns/PCC.jsp?topicid=102, accessed March 2004).

There have been warnings that, where conflicts arise between economic growth and the environment, 'corporate expertise' may seek ways to sideline environmental, social and ethical issues and stakeholders interests. The goal of business may be 'eco-efficiency', which has been defined as 'adding maximum value with minimum resource input and minimum environmental damage'. Some argue that companies may adopt bureaucratic, poorly transparent approaches that do not support the best practices or have scope for ongoing improvement. Increasingly popular corporate greening tools like environmental management systems may not be as beneficial as many claim. Some businesses are self-deluding, well-meaning but ineffectual; and, as discussed above, try greenwashing. However, a growing number genuinely try and usually have some degree of success.

There are signs of increasing green interest: business and law schools have been developing environmental courses. Green business interest is now spreading to developing countries – a recent review of progress has been provided by Utting (2004). For example, a large European car manufacturer recently established a professorship in sustainable development in a South African university. Corporate attitudes do seem to be shifting to more willing compliance, and sometimes to actually driving forward environmental management. While there are still many environmentalists who see business as 'the enemy', more and more now accept that the world is mainly run by commerce and that it must be worked with. Business 'greening' might be a key part of solving the world's environmental challenges. Cases like the 1984 Bhopal disaster and other commercial misdemeanours have helped convert many in business. There has also been a fair degree of self- prompted greening, plus encouragement by international bodies like the Business Council for Sustainable Development (Anon., 1993).

Some of the change toward environmental responsibility in business is being driven from 'within' by various stakeholders: sometimes it is management who have become enlightened and seek greening; staff may take the initiative and it can boost their company pride and morale; consumers may welcome or demand it; insurers may promote better environmental awareness to reduce the risk of accidents and costly claims; other companies, retailers or consumers may force it by refusing components or products from environmentally unsound companies; government or international regulations may prompt greening; NGOs, funding bodies and governments also encourage the shift (Buchholz, 1998). Already there are large companies which insist their component suppliers and other support subsidiaries meet strict environmental criteria. The crucial question is, does business just seek to comply with regulations, and avoid legal liability, taxation or insurance claims, or does it go beyond compliance? A tax per unit of pollution, or other environmental damage, does not discourage sudden discharges that can be difficult to monitor – it is better to adopt regulation that seeks to ensure that the capacity of the environment to cope with damage is not exceeded.

Money can be saved when businesses or other bodies practise recovery of waste products, and use less energy or raw materials. In practice, savings may be less clear cut, possibly recouped over very long periods or in ways that are not easy to measure (Schramm and Warford, 1989; Brown et al., 1992; United Nations Conference on Trade and Development [UNCTAD], 1993; Beaumont et al., 1993). The likelihood is that businesses will seek win–win approaches to try and reduce environmental damage and improve their competitive edge through improved productivity and/or lowered costs. For example, a company may issue marketing propaganda about its environmentally friendly paints, but does not tell customers these take some months to fully harden and are not as robust as the finishes they replace; the company may also save on paint-shop cleaning costs and on the cost of paint, but does not publicise this.

For developing countries, producing organic crops may be a chance to cut pollution and costly agrochemicals, and to break into otherwise tight export markets. Specialist crops or manufactured goods produced in an environmentally friendly manner may find a market in developed countries and attract enough profit for small-scale local producers to pay for transport and other costs, and still make a reasonable return. Much of the progress made in greening business is focused on the corporate sector, and care must be taken to prevent the small-business sector being overlooked. The manufacturing activity in developing countries is commonly undertaken by small-scale family businesses, shop-houses and workshops. These small enterprises are scattered amongst housing, difficult to monitor – indeed, are

Figure 10.1 *Residential and industrial mix. Shop houses and small factories with family accommodation above and behind, typical of much of South and Southeast Asia. Difficult to monitor and enforce regulations.*

Source: Author, 1978.

frequently illegal – and usually have little or nothing in the way of resources for environmental care, nor the ability to pay much if caught misbehaving (see Figure 10.1). The problem is not just urban and peri-urban manufacturing – in many rural areas of developing countries, small-scale mining operations are a problem. Small-scale commerce is little unionised, and often employs women and children and the elderly, groups which are especially vulnerable to workplace risks such as use of toxic compounds, inadequate safety clothing, solvent-based glues, welding equipment, cutting tools and so on. Business, whether small-scale or large, must care for the surrounding environment and the conditions its workers operate in; achieving that in developing countries can be a problem, as the Bhopal disaster showed. It helps if authorities adopt a precautionary principle approach.

Concluding points

- The world is increasingly globalised and affected more and more by business and consumerism. The greening of business will play a key role in future development. Indeed, with much of the globe's economic power in the hands of business, progress in environmental management will tend to be backed by commercial bodies.

- Some businesses have genuinely embraced green approaches, some have made half-hearted efforts, and others have cynically exploited greening, practising corporate 'greenwash'. Somehow, environmental managers must catalyse the change to green economics and green business practice through legislation, taxation, controls, propaganda and education.
- Consumerism and globalisation are seen by many environmentalists to threaten the drive for sustainable development. To counter them there are calls for new environment and development ethics; however, this is largely advocacy, with little clear statement of how it will actually be achieved. It does appear to be increasingly accepted that all countries are part of a globally interdependent environment.

Further reading

Cairncross, F. (1995) *Green Inc. Guide to Business and the Environment.* Earthscan, London. [A readable author with a particular gift for covering economic issues in a way anyone can grasp.]

Jacobs, M. (1991) *The Green Economy.* Pluto Press, London. [Lively and radical introduction to green economics.]

Turner, R.K., Pearce, D. and Bateman, I. (1993) *Environmental Economics: an elementary introduction.* Johns Hopkins University Press, Baltimore (MD). [Simple introduction to environmental economics.]

Scahltegger, S. and Burrit, R. (2002) *Contemporary Environmental Accounting: issues, concepts and practice.* Greenleaf Publishing, Sheffield. [An introduction to environmental accounting.]

The Ecologist (1995) *Whose Common Future? Reclaiming the Commons.* Earthscan, London. [Traces the degradation of environments as a consequence of enclosure of commons and the dispossession of local people.]

Welford, R. (2000) *Corporate Environmental Management: towards sustainable development.* Earthscan, London. [The third of three volumes on corporate environmental management, focusing on development issues and sustainability.]

Reed, D. (ed.) (1996) *Structural Adjustment, the Environment, and Sustainable Development.* Earthscan, London. [Explores the links between economic and social policies and the environment, particularly the impact of structural adjustment.]

Websites

World Bank Environment Department: http://www.worldbank.org, accessed January 2004

Green national accounting: http://www.oekpmomi.uio.no/memo/, accessed April 2004

Green Economics Resource Center: http://www.progress.org/baneker/home.htm, accessed April 2004

11 Environmental management and development: the future

Key chapter points

- This chapter reviews environmental management and development, and highlights key points. In particular, it argues that environmental managers must be on guard for misinterpretation – conclusions based on myth, poor data, received wisdom and inadequate knowledge. It is also crucial that they objectively select environmental management priorities.
- Environmentalists may hope for altered ethics – a shift of attitudes towards supporting sustainable development – rather than increasing consumerism and greed, but it would be wiser to assume, at best, a 'business-as-usual' scenario.
- The expectation is that the environment will remain stable and supportive to human development; that is unwise.

Earlier in this book the question was posed: what is environmental management? Is it the application of a set of tools? Is it a set of goals? Is it the activity of a cadre of administrators and consultants? Is it something everyone, even individual citizens, who use the environment and resources undertake? The simple answer is: all of these things. Ultimately, key development steering, starting and stopping has to be done – and that has to be overseen by environmental managers.

For issues like global environmental change, dealing with dangerous genetically modified organisms (GMOs), highly toxic and persistent chemicals and loss of biodiversity, management must be strategic and international. Much other environmental management is focused sectorally or at national, regional and local levels. When dealing with environmental management in developing countries it must be tuned to their needs – which is true for all countries. Much of the theory, tools, approaches, training and funding of current environmental management have evolved in developed countries; the process of adapting these to better suit developing country situations is incomplete, ongoing and often inadequate.

Developing countries are now generating approaches, tools and insights that are also very valuable to developed countries.

Politicians, media and development agencies frequently focus on 'hot spots' and 'hot topics'. There are dangers in this. With the former, there is a risk that inadequate data may have been misinterpreted as a reliable indication of a widespread problem (Thomas and Middleton, 1994; Fairhead and Leach, 1996; Leach and Mearns, 1996; Oates, 1999; Lomborg, 2001). When citizens, administrators and specialists concentrate on localities or specific problems that they perceive to be important, they can overlook other equally crucial things. When a localised problem is treated as a general issue, it could result in wasting resources or divert attention from more important but less obvious things. For example, spectacular gully erosion and badlands may attract remedial efforts at considerable cost and get limited returns, while a much more serious problem in the longer term is widespread gradual topsoil loss which gets little attention. The damaging topsoil loss will, if untreated, lead to serious vegetational change, water resource problems and difficulty growing crops, but it is difficult to recognise without careful monitoring until it is too late. Environmental management should alert authorities to such important, but less obvious, situations and ensure administrators look further than their immediate locality and issues attractive to the media. They should also establish whether developments are an early sign of a widespread problem, or a patchy and localised difficulty, and whether it merits attention.

Data collection and research may have a bias: areas near roads get seen by specialists who may hesitate to venture too far off the beaten track due to cost, time available, physical difficulty and dangers. Some areas are also easier for people to get access to, and thus the first, and perhaps the only areas, to show stress. Problem situations may also be identified because officials and researchers see them only at certain times when they appear degraded in a misleading way which is not typical of the longer term or wider scale.

One can recognise four challenging situations for environmental management, with many gradations between:

1   Resilient and insensitive environments: comparatively difficult to damage, recover easily.
2   Resilient but sensitive environments: comparatively easy to damage, but recover well.
3   Less resilient and insensitive environments: difficult to damage, recover badly.
4   Less resilient and sensitive environments: easy to damage and recover badly.

The first situation presents least difficulty and the fourth is the worst development situation. However, caution is needed; with current knowledge, environmental managers may have difficulty predicting how an ecosystem will respond to disturbance. Recovery after successive disturbances may not proceed in the same way. Resilience and vulnerability may also be unpredictable. Localities disrupted by civil unrest and warfare frequently suffer environmental problems as a consequence.

Priorities change – in the 1960s and 1970s those concerned with environmental matters worried about:

- *human population growth*
- *the threat of nuclear war*
- *the looming oil shortage 'energy crisis'.*

Today, climate change, especially human-induced global warming, has probably replaced these at the top of many people's agendas. Citizens, administrators, media and even specialists generally have a relatively limited attention span; established viewpoints, current paradigms and fashion influence what we worry about. I, and others, feel that global warming has taken up too much attention and diverted resources from equally important issues; Stott (2003) sees it as a 'hegemonic myth' resisting any criticism as misguided. Yet we cannot accurately or reliably predict what global environmental change is going to do and have little hope of safely controlling climate change; as a result, environmental managers should seek to make people and their developments more adaptable to any threat. The problem is that there has not been objective selection of environmental priorities. Attention focuses on what current opinion leaders or the ruling elite find worthy of attention and that may not be representative of a developing or developed country's views and needs.

Environmental managers must try to look at the whole picture and make objective judgements on what to prioritise; it may then be necessary to lobby to get it accepted. Often there are limited data and time available to do this, and it is not a precise art.

A vital role for environmental management is to help establish sensitive exploitation strategies. As already discussed, there are situations that are much less forgiving than others; this may be due to environmental causes or sometimes almost entirely for human reasons. In many situations problems arise through a mix of both environmental and human causation.

The status of 'easily damaged', perhaps immediately or possibly after some delay, and the unsatisfactory and insecure plight of people living in such a situation is often described as 'marginality'. While there is no precise definition, marginality is a condition in which people exploit resources in situations where even a slight shift in any one of many factors drastically alters the viability. Profitable exploitation may suddenly cease, leading to compensatory overuse – 'mining' the resource – and a failure of sustainability, environmental damage or abandonment. Profitable exploitation may suddenly become a possibility as a consequence of technical, economic, social or physical changes: a new road, a phase of above-average rainfall, a fashion for some natural resource, demand triggered by wartime shortages and so forth.

Those seeking a livelihood in marginal situations live 'near the edge' – outside developed nations there are few safety nets, so a failed harvest will probably not be cushioned by aid or government offers of alternatives. Environmental managers in a number of countries must cope with the problem of population movement into and population growth within marginal and vulnerable areas, and with increasing social differentiation, i.e. a widening gap between rich and poor (Glantz, 1994). Diverse factors cause marginalisation and social differentiation: economic development, innovations like new seeds, land speculation, eviction from conservation areas or enclosed land and many other causes. Marginalised people often cause worsening environmental damage as they struggle to survive, so environmental management frequently must deal with helping the poor toward sustainable livelihoods and away from causing and being victims of environmental degradation.

People may also be marginalised through environmental conditions; for example, poor soils, remote location, rugged terrain and arid conditions. These conditions are seldom static; there can be shifts to less favourable and more favourable conditions. Sometimes development takes place during a favourable phase and then people and environment suffer as conditions deteriorate – Western developers tend to see conditions in terms of 'average', and in many areas this is a dangerous viewpoint. Wildly fluctuating rainfall may give an 'average precipitation', but basing agricultural development plans on it would be foolish.

People can become marginalised for socio-economic reasons; typically the poor are the victims. One might debate whether marginality results from poverty or vice versa. Simply being different in some way from the dominant social group can result in marginalisation – and, as in Rwanda, it need not be a small proportion of the total. Various authors have explored why the 'poor stay poor' and how those near the margins can easily fall into a vicious spiral of worsening plight (Lipton, 1977; Blaikie and Brookfield, 1987). An example of this is people in an area with erratic precipitation who are weakened by a poor harvest, plant less as a consequence, and become more vulnerable. Living near the margin makes people vulnerable to physical and socio-economic setbacks. Modern societies have tended to degrade traditional coping strategies, exacerbating things. Marginalisation and vulnerability are complex and diverse processes which some have tried to understand and manage through a political economy approach (Blaikie, 1985).

In the past, peoples generally evolved livelihood strategies that incorporated alternative options and various hardship strategies – they made their own safety nets, especially in marginal situations. People who were not good at this failed to prosper and survive. The last few decades have seen widespread breakdown of strategies which have long worked in marginal environments. Sometimes this has been due to civil unrest, or legislation and restrictions, e.g. enforcement of borders and fencing of common land, which prevent mobility or access to alternative resources. There are situations where social change reduces the capacity of people to support each other in time of need, referred to as a breakdown of social capital; sometimes there are attitudinal shifts which degrade coping strategies, such as the rejection of taboos and superstitions, which help to ensure sustainable resource use by frightening people from exploiting some localities where stock can recover. The activities of government extension services, education, missionaries or any one of many other triggers can cause people to abandon traditional ways and workable lifestyles. Sometimes authorities seek to crush traditional ways as 'backward' and 'primitive' – even if they work better in harsh and unforgiving situations than modern 'alternatives'. Marginal land traditions have been a productive source of techniques for environmental managers presently seeking to improve security of livelihood, sustainability, adaptability and resilience (Messerschmidt, 1990). Much modern agriculture has been researched and developed under the opposite of marginal conditions on research stations – where the soil is ideal, water supplies unproblematic and communications and labour availability are good – yet most of the world is much nearer to marginal conditions than this.

Whether drylands, forest or wetlands, encroachment on marginal lands and various development pressures have disrupted established livelihood strategies. For example, in Iraq, the Marsh Arabs suffered when rivers supplying their marshlands were deliberately diverted; in parts of the Sahel, commercial agriculture and ranching has restricted traditional grazing areas; commercial logging has degraded forest areas. To cope with intrusions, indigenous groups have increasingly organised to gain firmer rights to their land and better control over their resources. There has also been a sharing of experiences between widely scattered indigenous groups and even mutual support. These peoples have a wealth of wisdom that is of value to others. Today, indigenous groups are increasingly being targeted for empowerment and participation in resource development, conservation and tourism in a way that would have been unheard of 50 years ago.

Regulating environmental activities

Some environmentally concerned non-governmental organisations (NGOs) have a huge worldwide membership and can command expertise, funds and legal support, which makes them a force to be reckoned with. Those tempted to undervalue the environment are now likely to face NGO opposition, some of it international and some indigenous – often both co-operate. NGOs and others are also able to call upon another new force which can pressure governments and large business over environmental issues – the media, including the Internet. In developed countries like The Netherlands, NGOs have played a significant role in negotiating covenants and agreements between business and the government which aim to reduce environmental damage. In developing countries, NGOs already act as 'whistle-blowers', support indigenous groups and those weakly-enfranchised, and have prompted better environmental awareness. NGOs like Greenpeace and Friends of the Earth gather data and lobby on issues like global climate change, tropical forest loss and other issues.

In developing countries and some developed countries government environmental agencies often lack willingness or 'teeth' to enforce good practice. More developed countries like the United States of America (USA), which has the Environmental Protection Agency with power to enforce environmental management decisions, have a fair degree of political freedom, openness of access to information and control over corruption. Many developing countries lack such things, and some have a civil service that is not politically neutral, stable and willing to act for the good of the whole nation – even some richer states are finding it difficult to maintain such valuable institutions.

Western environmentalists should not assume that environmental negotiations are undertaken with the interests of all held in reasonable regard. In the absence of world governance and adequate international law, countries have to rely on international bodies like the United Nations Environment Programme (UNEP) to temper national self-interest. Alas, the idea of siting the UNEP in Nairobi, to break away from the usual locations in Geneva, New York, Paris and other developed country cities, has meant it lies off the beaten track, which tends to hinder it. The UNEP also lacks funds, and visitors have to pay more to get to it. To some extent, some of the role of the UNEP has been taken on by other bodies like the United Nations Development Programme (UNDP), and by funding agencies like the World Bank and others.

There are two broad approaches to achieve environmental controls: regulation (command, the 'stick', or punishment) and incentives (reward, or 'carrot'). Strict adherence to 'the polluter pays principle' means less support for incentives and more of a command approach. Examples of green incentives include reduced landing fees for less noisy aircraft, cheaper unleaded petrol to discourage use of fuel with additives and subsidies for alcohol-fuelled cars.

Regulations can tell governments and business what they can and cannot do in detail, or can set less precise goals. Authorities can monitor pollution at the point of discharge – 'end-of-pipe' – or watch the state of the environment after emissions. The more regulations are agreed with those who have the potential to damage the environment, the less it is likely to cost a government. Regulations must be rigid enough to avoid enforcement problems – especially costly legal arguments – and to prevent evasion, and they must also work in practice.

One problem is that if an authority forces a company or service provider to meet a standard, the product or service costs are likely to rise and the customers tend to blame the company or service provider, rather than the authority. Business and services with old infrastructures are likely to have to pay increasing costs to meet standards, whereas a newly founded establishment is likely to select equipment that meets the standards and, thereby, incurs lower costs – in practice regulations can be unfair. Also, some activities are environmentally damaging and others less so – the risk is that costs of control will not be spread widely or fairly. Similarly, small companies are less likely to be able to meet standards than large, because they have fewer resources. Business may be tempted to discharge as much as a regulation allows, rather than seeking the best practical reduction; incentives may prompt a better response.

The risk of co-operating too closely with potential destroyers of the environment is that standards will not improve over time. Looking from the standpoint of the potential polluter – business or a developing country government – caution is needed to ensure regulations do not stifle industry and employment opportunities or cost too much. One way forward is to pursue a policy like that adopted in the United Kingdom (UK) since the 1990s: seeking the best available technique not entailing excess cost (BATNEEC). In the field of conservation the environmental protection versus employment conflict has often surfaced – in developed as well as developing countries – such as the logging workers' opposition to old forest protection in north-western USA.

Environmental management is in some ways similar to politics: it is 'the art of making good decisions based on insufficient reliable evidence.' Environmental management can be influenced by a range of standpoints, including:

- *High-tech, non-green to light-green environmentalism*; reliance on technology to achieve goals. Some argue that this is a 'quick-fix', clumsy and careless route, also that it is essentially working in conflict with the environment, trying to impose solutions, rather than working with nature.
- *Rational, mid-green, relatively shallow environmentalism*: strategies that are essentially environmentally sensitive. The approach does not exclude the use of technology but is more cautious than the previous stance. In a crowded world prone to sudden change, careful use of technology will be crucial.
- *Romantic, dark-green or deep environmentalism*: approaches which demand a drastic remodelling of society and ethics. Current attitudes are seen to be wrong and in conflict with the environment; in some cases mystical or religious change is advocated. Supporters are likely to reject technological solutions.

One can also recognise two extreme positions:

- *Ecocentric*: whatever the approach, the natural environment is given at least equal consideration with human needs.
- *Anthropocentric*: human needs have priority over nature.

There is of course, a huge diversity of other influences on environmental management: philosophical, religious, political, cultural, gender, etc. All approaches have faults: for example, a high-tech approach depends on ongoing breakthroughs in technology, whereas in reality there are often plateaux during which problems increase and solutions or funds to invest in new technology lag. Romantic and ecocentric approaches are all very well, but in a world with a large and growing population, the abandonment of technology or a failure to address human development needs would cause huge misery and unrest – small groups of people can live at one with nature with simple technology, but, with six-and-a-half billion people to feed, the environment somehow has to be stretched.

In a finite, crowded world the goal is to operate with minimal negative environmental impacts, provide sufficient concrete results fast enough to win popular support, develop strategies that take off under a wide range of often extremely unfavourable conditions, establish sustainable development and achieve integration of physical, social and economic planning and management. Sustainable development will have to be pursued at a manageable scale in a huge diversity of situations.

There is ongoing and wide-ranging discussion on the meaning of sustainable development and there is no universally accepted definition. Definitions are vague enough to be interpreted in various ways. Because sustainable development is so imprecise, those seeking clear benchmarks, firm concepts and reliable techniques may reject it. The vagueness also allows bodies to treat it as a 'flag of convenience', 'Trojan horse', or simply to pay it lip-service – failure to seriously embrace the concept (Lele, 1991; Costanza and Patten, 1995; Sneddon, 2002; Christie and Warburton, 2001).

Some see sustainable development as a principle and 'handrail' to guide development; like liberty or justice – poorly defined but vital for civilisation and progress. Others see sustainable development as a way of integrating diverse disciplines and activities or specialists to achieve common goals. More and more go further to seek practical applications, tools and workable strategies, and often these are environmental managers. There will never be one universal route to sustainable development: sometimes it will have to be 'bottom–up' – driven by the people, mainly via participatory, local-focused and low-tech means (Kapoor, 2001); elsewhere, perhaps overlapping, there may be 'top–down', high-tech initiatives. I have avoided using terms like 'appropriate', because hopefully all approaches will be appropriate, regardless of their management style or degree of local participation.

The driving force(s) behind sustainable development may be NGOs, government, business, international agencies, local people or specialist consultants. In some cases more than one of these may form a consortium, and where different initiatives overlap, adjoin or just have things to share, there will have to be good communication and co-ordination. Numerous sustainable development and environmental management networks have been established by the business community, NGOs, academics, aid agencies and so forth (International Network for Environmental Management, available online at http://gemi.org; Global Environmental Management Initiative, online at http://www.inem.org, accessed January 2004).

A spatial framework for environmental management and development is not a new idea; there have been numerous experiments since the 1930s, including the seminal work of the Tennessee Valley Authority. Most developed and developing countries have regional development planning and management bodies. Regional improvement, integration of conservation and other land uses, co-ordination of complex environmental issues that demand comprehensive or integrated planning and management, coping with problems of marginality, regional problem-solving, addressing unusual ecosystems development – all have attracted a regional approach (Friedmann and Weaver, 1979; Sale, 1985; Welford, 1996b; Brunckhorst, 2000). Some of the roots of ecodevelopment and later sustainable development lie in environmentally aware regional planning.

The World Bank (2003: 183) outlined 'pathways to sustainable development', focusing on the role of institutions, arguing that policies, institutions and the distribution of assets are interrelated (see Figure 11.1), and suggesting that that interrelationship can be managed and steered. Reviewing the future of environmental management, Wilson and Bryant (1997) concluded it would become more vital in development and would take place at multiple levels in the planning and implementation process.

What 'motors' will drive better environmental management and livelihood provision? A number of critics have pointed out that development is often too inflexible and too large: consequently the engineering is unadaptable, the financing

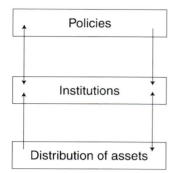

Policies shape institutions
and distribution of assets

Figure 11.1 *Policy–institutions–assets linkages.*

Source: Redrawn and modified from World Bank (2003) p. 183.

Distribution of assets
shapes institutions and policies

clumsy and the management slow to respond to challenges. Some see small-scale initiatives to be a better alternative, particularly since Schumacher (1973). One should not forget that small businesses can sometimes grow from garden-shed scale to huge multinational companies. Also, large companies and agencies can create sub-units which can attend to specialist activities. Small businesses may be good 'vehicles for adaptation', but there are other routes to adaptability, flexibility and diversification (Fuller, 2000). Hannam (1999) provides an interesting paper, focusing on the Indian Forest Service, to examine a crucial issue: how do the lead agencies involved in environmental management respond to challenges and adapt?

There are a number of key questions one might ask about environmental management:

* Can environmental management help poor countries by enhancing development?
* Is sustainable development possible without environmental management?
* What constitutes successful environmental management?
* What environmental management approach offers most promise?

This book has tried to explore these questions. Interestingly, only one of the eight Millennium Development Goals (number 7) actually pays any real attention to environmental management, and is at best vague about it (see Box 11.1 later in this chapter). There is clearly a long way to go before environmental management is adequately established.

Since the late 1970s, Kerala State (India) has been widely acclaimed as a model for developing countries, with lessons to pass on for those interested in environmental management, sustainable development and social development. Recent assessments of the Kerala example acknowledge that there has been good progress, but suggest that a rather rose-tinted view has been presented. Véron (2001) argues that it has not been as successful as was hoped, but recently a 'new' model of development has emerged in Kerala, which promises to better integrate sustainable development goals and improved environmental management. This new model is more decentralised and encourages participatory planning and collaboration between state, NGOs and civil bodies. This model should be of interest to environmental managers dealing with developing countries.

The environmental manager often acts like an artist, selecting from a palette of strategies, techniques, funding sources, insurance measures and whatever else is required, to create a strategy that best suits a situation. Then it is important to ensure the strategy is co-ordinated so as to minimise conflict with other efforts and maximise mutual benefits like training, co-operation, shared research and backup in event of disaster or problems. Debates about relative merits of low-tech versus high-tech, light-green versus dark-green approaches are probably wasted effort, because the struggle to promote sustainable development in a real world will almost certainly demand whatever means provide sustainable ends. Rather than recognise different routes, many now attack sustainable development strategies that differ from their choice; a task for environmental management will be to build bridges and motivate. Many of those working for sustainable development will be engrossed in local details or discipline concerns, so environmental managers will have to orchestrate the various efforts, referee trade-offs and maintain an overview from local up to global level.

Many companies, and probably the bulk of developed and developing country ruling classes, expect a future that will be broadly similar to today – 'business as usual'. The expectation is that there will be no major shift in commercial practices, little-altered access to resources, continuing domination by 'Western democracy', relative global peace and so forth – a rosy picture that discourages sufficient effort to seek sustainable development (Welford, 2000). In 1992, the UN World Conference on Environment and Development seemed to offer some hope – at least it established regular review meetings. Around the millennium there were some stocktaking exercises, scenario predictions, warnings and advocacy, but little serious, concrete forward thinking. That needs to be remedied.

Many of the world's problems are transboundary in nature; environmental management has to nurture and steer stakeholders toward international agreements. The efforts since the mid 1980s to control stratospheric ozone loss, via the Montreal Protocol on Emission of Ozone-depleting Chlorofluorocarbons, have been relatively successful. But the number of manufacturers of ozone-destroying chemicals is relatively small and alternative technology and controls have proved to be negotiable. Response to the perceived threat of global warming has been much less encouraging. Co-operation for environmental research, monitoring and forecasting has progressed, but treaties, which will cost nations large sums of money, perhaps even restrict their development, are difficult to agree. At the time of writing the Kyoto Agreement looked set to fail because the world's greatest carbon emitter, the USA, refused to sign up; Russia, another major contributor to greenhouse warming, has delayed signing; and India, China and other developed countries are set to pump out more and more carbon and methane (May et al., 1996; Lopez, 1999; Barrett, 2003).

Anthropogenic environmental change is by no means the only threat environmental managers should worry about: soil erosion, water shortages, global pollution, biodiversity loss, genetic engineering accidents and biological, chemical and nuclear weapons are but a few of the possible threats. There is a need for governments and citizens to be more aware of, and to treat seriously, the many natural threats which are currently largely ignored: large volcanic eruptions, mega-tsunamis, strikes by near-Earth bodies (meteorites, asteroids, comets and other heavenly bodies), geomagnetic changes, natural climate changes, extreme weather events, and many others.

Many people assume that twenty-first century humans are increasingly immune from natural disasters; however, the probability is that in spite of modern technology they are more vulnerable than ever before, because there are greater populations than in the past which depend on more complex institutions and supply chains. Also, people are nowadays often less adaptive and not as able to cope with stress as their ancestors. Assuming this assessment is correct, environmental managers should examine the vulnerability of various countries and groups of people, crucial environmental features and important resources, especially those which are irreplaceable.

Environmental knowledge is far from complete, and vulnerability is also affected by social and economic factors which are not easy to forecast, so it will never be possible to foresee or avoid all disasters. If adequate attention and expenditure is directed at vulnerability reduction, there is a better chance of achieving sustainable development.

Food is a key resource, and the poor in a number of developing countries continue to suffer shortages and access problems. Agricultural improvements since the 1960s have boosted crop yields, but there has been less improvement of sustainability and access or food reserves. In some regions where agriculture has been successfully improved over the last few decades there are now signs that fertilisers, pesticides and other crucial inputs are causing serious environmental problems. So there will have to be a 'Doubly Green Revolution', leading to sustainably raised yields with less environmental damage.

Disease control is another problem because in developing countries there are now more large cities than ever before, and many have rapidly increasing populations and infrastructure in a state of breakdown. Developed countries draw much of their food from the developing nations, and can be vulnerable to civil unrest in the latter. With modern air travel an infectious disease outbreak anywhere could quickly become a global threat. Rich nations need to support the poor, not just to satisfy some sense of charity, but also simply to enhance their own survival chances.

Crucial resources and expertise should be backed up in as many places and ways as possible, so that recovery from human unrest, accidents and natural disasters is more certain and easier. Developing vulnerability reduction measures demands close co-operation between environmental scientists, social scientists and other disciplines; sustainable development depends on social institutions as much as resources. Since the tragedy in New York on 9 September 2001, it has become clear how interdependent societies are and how easily resource supplies can be interrupted. Terrorism in the early twenty-first century has shown that unrest is difficult to externalise and that, no matter how powerful states are, there comes a point where they have to examine root causes of discontent and address them. The problems of developing countries, the global environment, trade and resources are all intertwined. Consequently it is very much in the interest of developed countries to help resolve environmental management problems in developing countries, because they share a common global environment and economy.

While the Cold War has ended and the threat of nuclear exchange between the great powers seems to have diminished a little, a number of developing countries have acquired or are working to obtain atomic weapons and other weapons of mass destruction. One recently rumoured development is 'gamma-ray weapons'. These are nuclear isomer weapons, which might be based on hafnium-178, allowing a warhead of only a few grams to yield the equivalent of 100 kg of high explosive, plus a lethal burst of radiation. If the rumours prove correct, and the weapons are easier to acquire than atomic bombs, the world will become more insecure than ever before. Terrorism appears to present new threats even without such developments. Fears about weapons of mass destruction played a part in triggering the 2003 Iraq conflict. There is a risk that in coming decades some crucial resources will cause disputes, when full attention and expenditure should be being directed to deal peacefully with environmental problems.

The conflicts and terrorism, human rights and environmental problems of the last few decades indicate a need for international bodies that are more effective than those currently established. The UN in general, and agencies like the UNEP in particular, need rethinking; they need adequate resourcing, improved administration, more 'teeth' to enforce policies and better representation of all countries, and they must adopt a more proactive approach. The future will probably see industrial and technological advancement in China and India, especially the latter. With growing 'sinobalisation', environmental negotiations and strategies will have to work for non-Western citizens.

Some international agencies spend too much on bureaucratic trappings and perks for their senior staff and have budgets that are too limited to achieve much. Also, too few people are willing to invest in combating environmental problems, especially those that do not directly affect them in the near future. Environmental management must encourage such investment.

Ironically, while the Western democratic approach looks unpromising to many as a route to stable world development and a sustainable environment, it has been the birthplace for many of the most effective environmental NGOs. Currently some of the most powerful environmental bodies are NGOs, and many have roots in humble private societies but have grown to the point where business and national governments have to heed their demands. Most NGOs have very dedicated staff and some are positively messianic in their approach; not surprisingly, they may sometimes eschew democratic routes, sound scientific methods or even basic ethics and can come into conflict with overall environmental or social well-being. Environmental management may have to moderate and steer such groups, especially in the absence of adequate international bodies.

The process of NGO formation has spread to developing countries and when environmentalists despair of finding a way to shift current trends toward better environmental care they might be encouraged by history: a small group of seventeenth-century English activists coping with dirt roads, weak law and order, and in some cases from opposite sides in a recent bloody civil war, managed to come together and found the Royal Society, which laid a good part of the foundations of modern science and western rationalism. With the Internet and increasingly robust and cheaper computers, there is a promising chance new groups can grow in poor countries and affect how the environment is cared for.

Future scenarios: forecasts and predictions

Looking to the future, it is possible to set goals and identify challenges. The Council of the European Union has published a set of goals shaped by its Millennium Development Goals – 'a Global Deal' (see Box 11.1). The central thrust of these is to eliminate poverty worldwide and to seek more sustainable development via aid, debt forgiveness, improved market access for developing countries and encouragement of 'better governance.' The European Union (EU) has also acknowledged the need to look further ahead – 20 years or more. Other bodies and agencies like the United Kingdom (UK) Department for International Development (DFID) have also recently published broadly similar 'millennium development goals'.

Box 11.1

The Millennium Development Goals (1990–2025)

1   Eradicate extreme poverty and hunger
    • Halve the proportion of people with less than US$1.0 a day
    • Halve the proportion of people who suffer from hunger
2   Achieve universal primary education
    • Ensure that boys and girls alike complete primary schooling
3   Promote gender equality and empower women
    • Eliminate gender disparity at all levels of education
4   Reduce child mortality
    • Reduce by two-thirds the under-five mortality rate
5   Improve maternal health
    • Reduce by three-quarters the maternal mortality ratio
6   Combat human immunodeficiency virus/acquired immune deficiency syndrome (HIV/AIDS), malaria and other diseases
    • Reverse the spread of HIV/AIDS
7   Ensure environmental sustainability
    • Integrate sustainable development into country policies and reverse loss of environmental resources
    • Halve the proportion of people without access to potable water
    • Significantly improve the lives of at least 100 million slum dwellers
8   Develop a global partnership for development
    • Raise official development assistance
    • Expand market access
    • Encourage debt sustainability.

Source: www.developmentgoals.org accessed October 2003.

Although the field of impact assessment has emerged since the 1970s to help predict environmental and social consequences of development, there is still far too little effort worldwide to identify future scenarios and plan ahead. Good forward planning allows contingency measures to be developed and possibly avoidance or vulnerability reduction activities; it can also allow costs to be spread and made more affordable. Some threats are easily and accurately predictable, others are a challenge, and there will never be wholly satisfactory awareness and preparation – nature and human behaviour will spring surprises. Environmental managers need to remind themselves of this; there is a limit to how much money and effort should be spent on avoidance and mitigation. Perhaps global warming activities have already drawn too much attention and expertise? Environmental management should focus on identifying ways in which people are vulnerable, crucial activities, infrastructure, and so on; because not all problems can be foreseen, a key need is to promote adaptability. Modern peoples and activities are generally less adaptable than was the case in the past. In developing countries, growing population, poverty and stretched infrastructures are increasing vulnerability as well as causing environmental degradation, which further exacerbates things. Improved adaptability and reduced vulnerability are a help even in the face of unexpected disasters.

The prediction of environmental, social and economic changes is not the only challenge: many administrators and citizens are not interested, or cannot afford to pay much attention. The reality is that few governments and many international bodies look further than about five years ahead. Environmental managers must try to expand the planning horizon. Unfortunately, looking ahead and spending may not yield results until long after politicians, planners and other experts have moved on. The trade-off of spending now for future return and expenditure, which may benefit people who did not pay, can be difficult to sell.

The quality of the environment and its component resources are held in trust by today's people for future generations. The concept of sustainable development stresses this, but short-term gain and individual or corporate profit presently predominate. Often citizens are aware of the consequences of such poor stewardship, but they cannot afford, especially in poor countries, to behave otherwise. Aid agencies, international funding bodies, NGOs and business are increasingly aware of these issues but finding adequate funds to assist people to take better care of the environment is a challenge.

Conflict and armaments waste money and frequently damage the environment: the Gulf War threatened to cause global environmental change through torched oil wells, and defoliation in Vietnam is still having marked impacts three decades later. Failure to resolve resource sharing and environmental problems have the potential to cause warfare. Little more than a hundred years ago few people in the UK or the USA would have imagined the widespread acceptance of payment of income taxes, purchase taxes, airport tax and so forth. Now, such levies are commonplace in developed countries, often levied at more than 30 per cent of gross income. There is hope that people can accept taxation for investing in the global environment and in assisting the poor; it should be possible to levy taxes on richer citizens and business for environmental management worldwide – even a tax of 1 or 2 per cent on developed country incomes would give huge revenue and offer more security and chance of sustainable development for all nations. If advertising can convince people to invest significant portions of their income in pointless and even harmful activities, it surely has the potential to 'sell' such global taxes.

There are environmental management issues that have generated dialogue and agreements between developed and developing countries. Global warming concern has prompted the formation of an international research, monitoring and discussion forum, the Inter-governmental Panel on Climate Change (IPCC), founded 1988. Concerned nations – often relatively weak developing states – have responded to warnings of likely sea-level change to form a lobby group, the Association of Small Island States (AOSIS). A catalytic role in the formation of AOSIS was played by NGOs, especially in 1995 during negotiations in Berlin. Threats like soil degradation, desertification and biodiversity loss have triggered less effective responses.

Some have argued that current approaches to environmental management are too technocratic, difficult to adapt and oriented to assisting state officials (Bryant and Wilson, 1998). There is a need to develop practical ways to achieve broadened-out environmental management, to enable the participation of non-state actors – farmers, herders and indigenous people – and to make it more adaptive (Kapoor, 2001; Norton and Steinemann, 2001). Technology can assist in the form of expert systems: these are computer programs that enable users to effectively perform tasks usually undertaken by scarce experts (for a review of expert systems application to environmental management see Warwick et al., 1993).

Recently there has been growing interest in past and ongoing environmental change. Some of this has been presented in a form that is accessible to the public, and which might help prompt interest in longer-term issues and the need to invest in managing such things. Before the late 1980s, relatively few researchers explored environment–human relationships: possibly this was a reaction to the excesses of environmental determinists before the 1940s, or perhaps it was because physical and social sciences had a strong mutual suspicion and different language. The linkage of environmental events to human fortunes has also been stimulated by severe El Niño/Southern Oscillation (ENSO) events of the 1980s and 1990s (Barrow, 2003).

Physical and social sciences seem more willing and able to communicate nowadays, and in the last 20 years or so, a growing proportion of those active in environmental management have come from the social sciences; indeed, they probably now exceed 50 per cent. More efforts are being made to understand how people perceive and interact with the environment and its other users. These insights are vital for 'managing'. In spite of the fashion for participation, empowerment and sustainable livelihoods, a lot of developments still have top–down management and are controlled by technocratic staff. And, whatever the approach – top–down or bottom–up – it is crucial to have adequate knowledge of environmental structure and function. Too often effective management is limited by gaps in knowledge and inadequate databases. In developing countries, if staff are city folk, Western-trained and unfamiliar with the local people and extra-urban conditions, there are likely to be problems. Some hope that distance learning may help overcome shortages of expertise; however, it may also discourage some who would leave their familiar surroundings and acquire a fresh outlook.

Developing countries should focus more on environmental stewardship adapted to their needs. Land husbandry, soil and water conservation and the sustainable livelihoods framework can help encourage more sensitive stewardship. As already stressed, there is no universal approach: strategies and methods must suit local needs, challenges and limitations and different ways may overlap and sometimes aid one another. However, transparency – making it possible for citizens and outside observers to see what is happening – is a universal need.

The globalisation of trade, communications, cultures and manufacturing are also influencing environmental management and stewardship (for a study of the 'ecology of globalisation' see French, 2000). Benedick (1999) argued that the future will be qualitatively different – environmental problems will increasingly be transboundary or global rather than the local or regional issues developers are currently familiar with. So co-ordinated global problem-solving therefore must have better support. In late 2003, UK environmentalist George Monbiot, known for his opposition to globalisation and multinational corporations, suggested that a World Parliament should be established, with representatives elected by blocks of 10 million people. Whether or not this is done through the UN, this might be a step toward giving the Earth's population more of a voice and a route to better international discussion and, hopefully, agreement.

Concluding points

- Environmental management must identify priorities, bottlenecks and barriers to counter degradation and, if possible, to achieve sustainable development.
- There are many strategies for improving environmental management and pursuing sustainable development. Environmental managers should take the initiative and steer the development and implementation of these and ensure co-ordination and regular review when they are running.
- Ways have to be found to 'broaden out' environmental management to enable and encourage wider participation.
- There are many problems and threats that need attention. Environmental managers must watch for and warn developers of these, and encourage them to focus much more on improving human adaptability and reducing human vulnerability.

Further reading

Bryant, R.L. and Wilson, G.A. (1998) Rethinking Environmental Management. *Progress in Human Geography* 22(3), 321–43. [An assessment of the state of environmental management.]

Wilson, G.A. and Bryant, R.L. (1997) *Environmental Management: new directions for the twenty-first century.* University College London Press, London. [Stimulating introduction, although it does not focus much on developing countries. Argues that environmental management is a multilayered process involving a wide range of actors.]

Dodds, F. (1997) *The Way Forward: Beyond 'Agenda 21'*. Earthscan, London. [Observations made soon after the Rio Earth Summit on how to seek sustainable development.]

Wade, R. (2003) *Governing the Environment: the World Bank and the struggle to redefine development.* The World Bank, Washington (DC). [Review of the way forward through World Bank eyes.]

Athanasiou, T. (1998) *Slow Reckoning: the ecology of a divided planet.* Vintage, London. [Radical call for better management of the environment and the relationship between rich and poor nations.]

Websites

The Rio Declaration: http:/www.un.org/documents/ga/conf151/aconf15126-1annex1.html

The Johannesburg World Summit on Sustainable Development 2003: http://www.johannesburgsummit.org/html/documents/summit_docs/plan_final1009.doc

The Johannesburg Declaration on Sustainable Development 2003: http://www.johannesburgsummit.org/html/documents/summit_docs/1009wssd_pol_declaratio.doc

(All accessed September 2003)

Glossary

Adaptation Adjustments in ecological, social or economic systems in response to actual or expected (perceived) change.

Agenda 21 A set of goals and proposals published following the 1992 Rio Earth Summit that promotes sustainable development. All countries have been encouraged to adapt the proposals and incorporate them in governance and management.

BATNEEC Best available technique not entailing extra cost.

BOD Biochemical oxygen demand. Water enriched with a pollutant(s) prompts excessive microbial activity (eutrophication) and algae growth and becomes depleted of oxygen and enriched with ammonia, leading to extensive kills of higher aquatic life forms. A body of water can only handle so much pollution before BOD and/or COD (see below) overwhelms it. Once seriously damaged, the water body can render much less pollution, if any, 'safe'. Nitrogen in sewage and other organic pollutants and phosphates from fertilisers or detergents are usually responsible for BOD.

BPEO Best possible environmental option.

BSE Bovine spongiform encephalopathy. A cattle 'disease' (there is still some debate about causes) which can affect humans if they consume contaminated material, ultimately causing severe mental deterioration.

Cassandras Those catastrophist and pessimistic environmentalists who preach environmental disaster through human misdevelopment The opposite are the **cornucopians** (see below). A cruder alternative for Cassandras is 'prophets of doom'.

COD Chemical oxygen demand. Pollutants cause chemical reactions, which deplete water of oxygen and kill aquatic life.

Cold War A period (roughly 1946 to 1986) of mutual distrust, power struggles, arms race and propaganda contest between the First and Second World.

Commodities Non-food agricultural produce, e.g. cotton, coffee, tea, rubber and forest products.

Common resource Open access to a number of users. The argument goes that overexploitation of common resources was prevented for centuries by disease and warfare, among other factors, but that the population has since overcome controls through developments such as better healthcare and law and order. The assumption has been that users maximise their exploitation, leading to a crash. However, many argue that there is seldom 'open access' to common resources – tradition, taboos and rules prevent overexploitation. Sometimes these controls may break down or be inadequate for modern demands. In recent years many common resources have been 'privatised' by commercial or state bodies, with little opposition because traditional users have no legal clout or documentation to resist. The dispossessed may move and degrade other marginal land or lose part of their subsistence.

Consumerism Hopes and demands for material possessions, typically luxury goods, which rapidly become 'obsolete' and get discarded. Consumerism is part-driven by fashion, advertising and the globalisation of Western culture.

Cornucopian Applied to optimists who feel (the implication is uncritically) that an environmental crisis will be overcome without too much trouble. These optimists have also been called 'prophets of boom' (see **Cassandras**, above).

Corporate social responsibility Voluntary approach to environmental management by business, as opposed to command and control by government. Business takes on some of the responsibility previously held by government.

Cultural ecology Study of the adaptation of human societies or populations to their environments, emphasising the management of technology, economy and social organisation, through which culture mediates.

Development A process which increases the capacity to meet people's needs and improve the quality of human life; conscious pursuit of a better quality of life for people.

Drought There are many definitions, none universal. The following is probably as good as any: a deficiency of precipitation from the expected norm, which means there is too little moisture to sustain agriculture, vegetation cover and often groundwater under existing demands. Often drought is seen to be caused by climatic fluctuations and to be of a temporary nature – that may not be the true situation. Drought is frequently (by no means the only) cause of hunger and famine.

Desertification (desiccation) Loss of vegetational productivity and degradation of plant cover and soil, such that less moisture infiltrates the soil, and possibly leading to regional deterioration in precipitation due to changed albedo and more wind exposure as vegetation deteriorates. Desertification is generally attributed directly or indirectly to poor soil and water management.

Earth Summit The 1992 United Nations Conference on Environment and Development (UNCED) held at Rio de Janeiro. This was the second major UN environment and development conference, the first being at Stockholm in 1972 and the third at Johannesburg in 2002.

Ecology Study of the relationships between living organisms and/or between living organisms and their environment. The term was probably first used by the

German naturalist Ernst Haeckel in 1866 to describe the study of the 'web' of organisms and environment.

Ecologism Deep-seated objections to the way that the environment is affected by society. Adherents are not content with correcting a problem: they seek to examine why it was generated and how social change, new outlooks or ethics can avoid it.

Ecotourism Environmentally sensitive tourism, which may or may not be based on nature, and which is characterised by investing significant amounts of the revenue it makes in environmental management.

EMAS European Union eco-management and audit system. Not to be confused with the eco-management and audit regulation (EMA). Originally the intention was to make the scheme obligatory; however, lobbying led to it becoming a voluntary but encouraged measure, which came into force in 1995 (Council Regulation of the European Community – EEC No. 1836/93).

Empowerment Process by which individuals, typically including the poorest, are assisted to take more control over their lives so that they become agents of their own development.

Environmental determinism The concept that human fortunes are predominantly caused by physical conditions and/or genetics and not through the exercise of free will, learning and so forth. The crude and simplistic determinism of the 1870s to 1940s is now rejected; however, there has been renewed interest in more cautious neo-environmental determinism since the late 1980s. This suggests that nature does impact on human development but does not determine the outcome.

ENSO El Niño/Southern Oscillation, a large-scale ocean climate phenomenon. Climatic conditions in the Pacific off eastern coast of south-east Asia alters ocean circulation, which in time weakens the upwelling of nutrient-rich cold deepwater off South America. The impact on climate and wildlife is marked. These El Niño events are recurrent but not quite predictable. The opposite of an El Niño is where there are strongly 'normal' conditions off South America, known as a La Niña. ENSO events have extended impact on a wide part of the world for some time after becoming manifest off South America.

Environmentalism Planned intervention to secure improvement in environmental quality. There is a huge diversity of types of environmentalism. While not a clear-cut division, one could separate 'green' activists from environmentalists, because the former are more politicised; however, both share a desire to protect and improve the environment.

Environmental possibilism The idea that environment plays some role in human fortunes, but does not determine the outcome: i.e. it limits and prompts but does not control.

Environment The sum total of conditions within which organisms live; the result of interaction between non-living (abiotic) physical and chemical components and, where present, those that are living (biotic).

Ethics Standards of proper conduct.

Expert system A computer program which assembles the knowledge of experts in some field and which can often learn and improve with use. It is used by relatively unskilled staff to make acceptable decisions, having first entered available information. For example, symptoms of a disease are fed in to the system to prompt a diagnosis for a healthcare worker unable to consult a doctor.

Famine Not a precise term, but essentially a food shortage sufficient to cause disastrous levels of debilitation and death. Fatalities are often through hunger weakening immunity to various diseases rather than starvation alone. Society and family behaviour is likely to breakdown (see **malnutrition** and **undernutrition**).

Gaia hypothesis Formulated by James Lovelock in 1979, this postulates that the biosphere operates as a single 'organism'. Complex checks and balances between the living and non-living maintain the Earth's environment – a homeostatic system. It was initially treated with scepticism, but there is now growing interest in the hypothesis. The implication for development, if the hypothesis is correct, is that humans must work carefully with earth systems – 'fit in'.

Gender-benders Slang term for compounds with an oestrogen-like effect. At low concentration, chemicals such as polychlorinated biphenyls (PCBs), dioxins, compounds used in plastics manufacture, some pesticides and contraceptive pill hormones in sewage can result in a region's fish and reptiles becoming predominantly female or neuter. There are also fears that human sperm counts have been lowered; however, Lomberg (2001) offers a possible behavioural cause as an alternative to pollution.

Global Environmental Facility A fund set up in haste in 1991 by rich developed countries to support developing nations that undertake works that promise global environmental gains. Managed by the World Bank, it has been accused of secretiveness. Also, it only funds activities promising global benefits. It is supposed to help developed countries deal with biodiversity, climate change, desertification and persistent pollutants.

Governance The means by which policies are implemented and monitored through administration, policy making and the rule of law. In short: the manner of governing.

Greenwash Misinformation disseminated by an organisation so as to present an environmentally responsible public image.

Hazard assessment There is often confusion with risk assessment. To illustrate the difference: imagine a person wishing to cross the Atlantic: the hazard is drowning, so the person has a choice of making the journey by rowing boat or ocean liner, which clearly present different risks, or drowning. Hazard assessment seeks to identify threats and their nature; risk assessment is also concerned with the probability, severity, frequency and timing of threats.

Heavy metals Often toxic at low concentrations, and may be concentrated by organisms near the base of the food-web so organisms near the top are seriously affected. Heavy metals include mercury, cadmium, zinc, lead, copper, chromium and silver. Domestic waste, human sewage and some livestock waste can be rich in heavy metals.

Barrow, C.J. (1997) *Environment and Social Impact Assessment: an introduction.* Arnold, London.

Barrow, C.J. (1999a) *Environmental Management: principles and practice.* Routledge, London.

Barrow, C.J. (1999b) *Alternative Irrigation: the promise of runoff agriculture.* Earthscan, London.

Barrow, C.J. (2000) *Social Impact Assessment: an introduction.* Arnold, London.

Barrow, C.J. (2003) *Environmental Change and Human Development: controlling nature?* Arnold, London.

Barry, J. (1999) *Environmental and Social Theory.* Routledge, London.

Bartelmus, P. (1986) *Environment and Development.* Allen and Unwin, London.

Bartelmus, P. and Seifert, E.K. (eds) (2003) *Green Accounting.* Ashgate, Abingdon.

Bate, J. (1991) *Romantic Ecology: Wordsworth and the environmental tradition.* Routledge, London.

Baumast, A. (2001) Environmental management: the European way. *Corporate Environmental Strategy* 8(2), 148–56.

Baxter, W., Ross, W.A. and Spaling, H. (2001) Improving the practice of cumulative effects assessment in Canada. *Impact Assessment and Project Appraisal* 19(2), 253–62.

Beaumont, J.R., Pedersen, L.M. and Whitaker, B.D. (1993) *Managing the Environment: business opportunity and responsibility.* Butterworth-Heinemann, Oxford.

Beeton, A.M. (2002) Large freshwater lakes: present state, trends, and future. *Environmental Conservation* 29(1), 21–38.

Benedick, R.E. (1999) Tomorrow's environment is global. *Futures* 31 (1999), 937–47.

Bennett, G. (1992) *Dilemmas: coping with environmental problems.* Earthscan, London.

Berger, P.L. (1987) *The Capitalist Revolution.* Wildwood House, Aldershot.

Berkes, F. (ed.) (1989) *Common Property Resources: ecology of community-based sustainable development.* Belhaven, London.

Berkes, F. (1999) *Sacred Ecology: traditional ecological knowledge and resource management.* Taylor and Francis, Ann Arbour (MD).

Berkes, F. and Folke, C. (eds) (1998) *Linking Social and Ecological Systems: management practices and social mechanisms.* Cambridge University Press, Cambridge.

Bernstein, H. (1973) *Underdevelopment and Development: the third world today.* Penguin, Harmondsworth.

Bernstein, H., Crow, B., Mackintosh, M. and Martin, C. (eds) (1990) *The Food Question: profits versus people?* Earthscan, London.

Bigg, T. (1995) The UN Commission on Sustainable Development: a non-governmental perspective. *Global Environmental Change* 5(3), 251–3.

Biswas, A.K. (2001) World Water Forum: in retrospect. *Water Policy* 3, 351–6.

Blaikie, P.M. (1985) *The Political Economy of Soil Erosion in Developing Countries.* Longman, Harlow.

Blaikie, P.M. and Brookfield, H. (eds) (1987) *Land Degradation and Society.* Methuen, London.

Blaikie, P.M., Cannon, T., Davis, I. and Wisner, B. (eds) (1994) *At Risk: natural hazards, people's vulnerability and disasters.* Routledge, London.

Blaney, G. (1982) *The Triumph of the Nomads: a history of ancient Australia.* Macmillan, Melbourne.

Bookchin, M. (1987) Social ecology versus deep ecology. A challenge for the ecology movement. *Green Perspectives* Nos 4 and 5.

Boons, F.A.A. and Baas, L.W. (1997) Types of industrial ecology: the problem of co-ordination. *Journal of Cleaner Production* 5(1–2), 79–86.

Borri, F. and Boccaletti, G. (1995) From total quality management to total quality environmental management. *The TQM Magazine* 7(5), 38–42.

Borrini-Feyerabend, G. (1996) *Collaborative Management of Protected Areas: tailoring the approach to the context.* IUCN, Gland.

Boserüp, E. (1965) *The Conditions of Agricultural Growth: the economics of agrarian change under population pressure.* Allen and Unwin, London.

Boserüp, E. (1981) *Population and Technology.* Blackwell, Oxford.

Boserüp, E. (1990) *Economic and Demographic Relationships in Development.* Johns Hopkins University Press, Baltimore (MD).

Boulding, K. (1971) The economics of the coming spaceship earth. *Development Digest* IX(1): 12–15

Boyce, J.K. (2002) *The Political Economy of the Environment.* Edward Elgar, Cheltenham.

Briassoulis, H. (2001) Sustainable development and its indicators: through a (planner's) glass darkly. *Journal of Environmental Planning and Management* 44(3), 409–27.

Brookfield, H.C. (1973) On one geography and a third world. *Institute of British Geographers Transactions* Old series No. 58: 1–20.

Brown, D.O. (1998) Debt-funded environmental swaps in Africa: vehicles for tourism development? *Journal of Sustainable Tourism* 6(1), 69–78.

Brown, K., Tompkins, E.I. and Adger, N. (2002) *Making Waves: integrating coastal conservation and development.* Earthscan, London.

Brown, L.R. Flavin, C. and Postel, S. (1992) *Saving the Planet: how to shape an environmentally sustainable global economy.* Earthscan, London.

Brunckhorst, D.J. (2000) *Bioregional Planning. Resource Management beyond the New Millennium.* OPA, Amsterdam.

Bryant, R.L. and Bailey, S. (1997) *Third World Political Ecology.* Routledge, London.

Bryant, R.L. and Wilson, G.A. (1998) Rethinking environmental management. *Progress in Human Geography* 22(3), 321–43.

Buchholz, R.A. (1998) *Principles of Environmental Management: the greening of business.* Pearson-Prentice Hall, London.

Bull, D. (1982) *A Growing Problem: pesticides and the Third World poor.* Oxfam, Oxford.

Bulloch, J. and Darwish, A. (1993) *Water Wars: coming conflicts in the Middle East.* Victor Gollancz, London.

Bunney, S. (1990) Prehistoric farming caused 'devastating' soil erosion. *New Scientist* 125(1705), 29.

Buttel, F.H. (1978) Environmental sociology: a new paradigm? *The American Sociologist* 13, 252–6.

Button, K.J. and Pearce, D.W. (1989) *Improving the Urban Environment: how to adjust national and local government policy for sustainable urban growth.* Pergamon, Oxford.

Cairncross, F. (1991) *Costing the Earth.* The Economist Books, London.

Cairncross, F. (1995) *Green Inc. Guide to Business and the Environment.* Earthscan, London.

Cairncross, S. and Feachem, R.G. (1983) *Environmental Health Engineering in the Tropics: an introductory text* (2nd edn). Wiley, Chichester.

Calder, I.R. (1999) *The Blue Revolution: land use and integrated water resource management.* Earthscan, London.

Caldwell, L.K. (1990) *International Environmental Policy: emergence and dimensions.* Duke University, Durham (NC).

Caldwell, M. (1977) *The Wealth of Some Nations.* Zed, London.

Capra, F. (1982) *The Turning Point: science, society and the rising culture.* Wildwood House, London.

Caratti, P., Dalkmann, H. and Jiliberto, R. (eds) (2004) *Analysing Strategic Environmental Assessment: toward better decision-making.* Edward Elgar, Cheltenham.

Carruthers, I. and Chambers, R. (1981) Rapid appraisal for rural development. *Agricultural Administration* 8: 407–22.

Carson, R. (1962) *Silent Spring.* Houghton Mifflin, St Louis (MI).

Carson, R. (2004) *Contingent valuation: a comprehensive bibliography and history.* Edward Elgar, Cheltenham.

Cascio, J. (2003) *ISO 14000 Guide: the new international environmental management standard.* McGraw-Hill, New York (NY).

Castro, F. (1993) *Tomorrow is too Late: development and the environmental crisis in the third world.* Ocean Press, Melbourne (USA: Talmun Co., New York [NY]).

Chambers, P.E., Jensen, R. and Whitehead, J.C. (1996) Debt-for-nature-swaps as noncooperative outcomes. *Ecological Economics* 19(2), 135–46.

Chambers, R. (1997) *Whose Reality Counts? Putting the first last.* Intermediate Technology Publications, London.

Chambers, R., Longhurst, R. and Pacey, A. (eds) (1981) *Seasonal Dimensions to Rural Poverty.* Frances Pinter, London.

Chander, P. and Khan, M.A. (2001) International treaties on trade and global pollution. *International Review of Economics and Finance* 10, 303–24.

Chandler, R. (1988) *Understanding the New Age.* Word, London.

Cheney, J. (1989) Postmodern environmental ethics: ethics as bioregional narrative. *Environmental Ethics* 11(2), 117–34.

Chiesura, A. and de Groot, R. (2003) Critical natural capital: a socio-cultural perspective. *Ecological Economics* 44(2–3), 219–31.

Chisholm, M. (1982) *Modern World Development: a geographical perspective.* Hutchinson, London.

Christie, I. and Warburton, D. (2001) *From Here to Sustainability: politics in the real world.* Earthscan, London.

Churchill, R. and Freestone, D. (eds) (1991) *International Law and Global Climate Change.* Graham & Trotman, London.

Cicin-Sain, B. and Knecht, R.W. (1998) *Integrated Coastal and Ocean Management.* Island Press, Washington (DC).

Clapp, J. (2002) What the pollution havens debate overlooks. *Global Environmental Politics* 2(2), 11–19.

Clark, J.R. (1996) *Coastal Zone Management Handbook.* Lewis Publishers, Boca Raton (FL).

Clarke, A.H. (2002) Understanding sustainable development in the context of other emergent environmental perspectives. *Policy Sciences* 35, 69–90.

Clarke, R. (1991) *Water: the international crisis.* Earthscan, London.

Cleary, D. (1990) *Anatomy of the Amazon Gold Rush.* Macmillan, London.

Cohen, J.E. (1996) *How Many People Can the Earth Support?* W.W. Norton, New York (NY).

Colby, M.E. (1991) Environmental management in development: the evolution of paradigms. *Ecological Economics* 3, 193–213.

Colombo, U. (2001) The Club of Rome and sustainable development. *Futures* 33(2001), 7–11.

Commoner, B. (1972) *The Closing Circle.* Bantam, New York (NY).

Conroy, C. and Litvinoff, M. (eds) (1988) *The Greening of Aid: sustainable livelihoods in practice.* Earthscan, London.

Consumers' Association of Penang (1984) *The Lessons of Bhopal: a community action resource manual.* CAP, P.O. Box 1045, Penang, Malaysia.

Conway, G. (1997) *The Doubly Green Revolution: food for all in the 21st century.* Penguin, London.

Cooper, C. (1981) *Economic Evaluation and Environment: a methodological discussion with particular reference to developing countries.* Hodder and Stoughton, London.

Cooper, C. (1990) *Green Christianity: caring for the whole creation.* Hodder and Stoughton, London.

Cooper, D.E. and Palmer, J.A. (1998) *Spirit of the Environment: religion, value and environmental concern.* Routledge, London.

Corbridge, S. (1991) Definitions of development. *Geographical Review* 5(2), 15–8.

Cosgrove, D. (1990) Environmental thought and action: pre-modern and post-modern. *Transactions of the Institute of British Geographers* new series 15(3), 344–58.

Costanza, R. (ed.) (1991) *Ecological Economics: the science and management of sustainability.* Columbia University Press, New York (NY).

Costanza, R. and Patten, B. (1995) Defining and predicting sustainability. *Ecological Economics* 15(3), 193–6.

Cotgrove, S. (1982) *Catastrophe or Cornucopia? The environment, politics and the future.* Wiley, Chichester.

Council on Environmental Quality and Department of State (1980) *The Global 2000 Report to the President: entering the twenty-first century.* US Government Printing Office, Washington (DC) (Penguin edn 1982).

Council on Environmental Quality and Department of State (1981) *Global Future: a time to act.* Council on Environmental Quality, Washington (DC).

Covello, V.T. and Merkhofer, M.W. (1993) *Risk Assessment Methods: approaches for assessing health and environmental risks.* Plenum Press, New York (NY).

Crognale, G. (1999) *Environmental Management Strategies: the 21st century perspective.* Pearson-Prentice Hall, London.

Crosby, A. (1987) *Ecological Imperialism: the biological expansion of Europe 900–1900.* Cambridge University Press, Cambridge.

Crosson, P.R. (1997) Will erosion threaten agricultural productivity? *Environment* 39(8), 4–31.

Crumley, C.L. (ed.) (1994) *Historical Ecology: cultural knowledge and changing landscapes.* University of Washington Press, Washington (DC).

Crump, A. (1991) *Dictionary of Environment and Development: people, places, ideas and organizations.* Earthscan, London.

Dale, T. and Carter, V.G. (1954) *Topsoil and Civilisation.* University of Oklahoma Press, Norman (OK).

Daly, H. (1992) Free market environmentalism: turning a servant into a bad master. *Critical Review* 6(2–3), 171–83.

Daly, H. and Townsend, K. (eds) (1993) *Valuing the Earth: economics, ecology, ethics.* MIT Press, Cambridge (MA).

Darwin, C. (1859) *The Origin of Species by Means of Natural Selection: or the preservation of favoured races in the struggle of life.* John Murray, London (Penguin edn 1968).

Dasgupta, P. (2001) *Human Well-being and the Natural Environment.* Oxford University Press, Oxford.

Davis, M. (2001) *Late Victorian Holocausts: El Niño famines and the making of the third world.* Verso, London.

De Bardeleben, J. (1986) *The Environment and Marxism–Leninism: the Soviet and East German experience.* Westview Press, Boulder (CO).

De LaPerriere, R.A.B. and Seuret, F. (2000) *Brave New Seeds: the threat of GM crops to farmers*. Zed, London.

Deng, H. (1992) Urban agriculture as urban food supply and environmental protection subsystems in China. Paper to the 1992 Conference on Planning for Sustainable Urban Development, University of Wales Cardiff, Cardiff.

Dery, D. (1997) Coping with 'latent time bombs' in public policy. *Environmental Impact Assessment Review* 17(5), 413–25.

Desai, M. (1995) Greening of the HHDI? in A. McGillivray (ed.) *Accounting for Change*. The New Economics Foundation, London, pp. 21–36.

Destra, A. (1999) *Environmentally Sustainable Economic Development*. Praeger, Westport (CT).

Devall, B. and Sessions, G. (1985) *Deep Ecology: living as if nature mattered*. Peregrine Smith, Salt Lake City (UT).

Devas, N. and Rakodi, C. (1993) *Managing Fast Growing Cities: new approaches to urban planning and management*. Longman, Harlow.

De Weerdt, S. (2004) Dark secret of the lake. *New Scientist* 197(2436), 40–3.

Dien, M.I. (2000) *The Environmental Dimensions of Islam*. Lutterworth Press, Cambridge.

Dobson, A.P. (1996) *Conservation and Biodiversity*. Scientific American Library, New York (NY).

Dorcas, R., Hewitt, T. and Harriss, J. (eds) (2000) *Managing Development: understanding inter-organisational relationships*. Sage, London.

Douglas, I. (1983) *The Urban Environment*. Edward Arnold, London.

Douglas, M. and Wildavsky, A. (1982) *Risk and Culture*. University of California Press, Berkeley (CA).

Downs, T. (2000) Challenging the culture of underdevelopment and unsustainability. *Journal of Environmental Planning and Management* 43(5), 601–22.

Drakakis-Smith, D.W. (1987)*The Third World City*. Methuen, London (2nd edn 2000, *Third World Cities*, Routledge, London).

Dunn, B.C. and Steinemann, A. (1998) Industrial ecology and sustainable communities. *Journal of Environmental Planning and Management* 48(6), 661–73.

Dupâquier, J. and Grebenik, E. (eds) (1983) *Malthus Past and Present*. Academic Press, London.

Eckersley, R. (1988) The road to ecotopia? Socialism versus environmentalism. *The Ecologist* 18(4/5), 142–8.

Eckersley, R. (ed.) (1996) *Markets, the State and the Environment: towards integration*. Macmillan, Basingstoke.

Eckholm, E.P. (1976) *Losing Ground: environmental stress and world food prospects*. Norton, New York (NY).

Ecologist, The (1995) *Whose Common Future? Reclaiming the commons*. Earthscan, London.

Eden, M.J. and Parry, J.T. (eds) (1996) *Land Degradation in the Tropics: environmental and policy issues*. Pinter, London.

Edwards, C.A., Lal, R., Madden, P., Miller, R.H. and House, G. (eds) (1990) *Sustainable Agricultural Systems*. Soil and Water Conservation Society, Ankeny (IO).

Ehrlich, P.R. (1962) *The Population Bomb*. Ballentine Books, New York (NY).

Ehrlich, P.R. and Ehrlich, A.H. (1990) *The Population Explosion*. Simon and Schuster, New York (NY).

Ehrlich, P.R., Ehrlich, A.H. and Holdren, J.P. (1970) *Ecoscience: population, resources, environment*. Freeman, San Francisco (CA).

Eisenberg, E. (1998) *The Ecology of Eden: nature and human response.* Knopf, New York (NY).

El-Kholy, O.A. (2001) Trends in environmental management in the last 40 years, in Ted Munn (ed.) *Encyclopedia of Global Environmental Change* (vol. 4). Wiley, Chichester, pp. 15–20.

Ellen, R. (1982) *Environment, Substance and Systems: the ecology of small scale social formations.* Cambridge University Press, Cambridge.

Elton, C.S. (1958) *The Ecology of Invasions by Animals and Plants.* Chapman and Hall, London.

Endres, A. and Finus, M. (2002) Quotas may beat taxes in a global emission game. *International Tax and Public Finance* 9(6), 707–21.

Epps, J.M. (1996) Environmental management in mining: an international perspective of an increasing global industry. *Journal of the South African Institute of Mining and Metalurgy* 96(2), 67–70.

Erickson, S.L. and King, B.J. (1999) *Fundamentals of Environmental Management.* Wiley, New York (NY).

Erkman, S. (1997) Industrial ecology: an historical view, *Journal of Cleaner Production* 5(1-2), 1–10.

Erocal, D. (ed.) (1991) *Environmental Management in Developing Countries.* OECD Press, Paris.

Ervin, D.E. and Shmitz, A. (1996) A new era of environmental management in agriculture? *American Journal of Agricultural Economics* 78(5), 1198–206.

Escobar, A. (1992) Reflections on development: grassroots approaches and alternative policies in the third world. *Futures* 24 (1992), 411–36.

European Commission (1998) *Guidelines for the Application of the Precautionary Principle.* Directorate General XXIV (Consumer Policy and Consumer Health Protection). European Commission, Brussels.

European Commission (2000) Common position 2000/25/EC adopted by the Council on 30 March 2000. *Official Journal of the European Communities* C137, 11–22.

Evanari, M., Shanan, L. and Tadmore, N. (1982) *The Negev: the challenge of a desert.* Harvard University Press, Cambridge (MA).

Evans, D. (1991) *A History of Nature Conservation in Britain.* Routledge, London.

Fagan, B. (2000) *Floods, Famines and Emperors: El Niño and the fate of civilizations.* Pimlico Press, London.

Fairhead, J. and Leach, M. (1996) *Misreading the African Landscape: society and ecology in a forest-savanna mosaic.* Cambridge University Press, Cambridge.

FAO (2001a) *The State of Food and Agriculture 2001.* Food and Agriculture Organization of the UN, Rome.

FAO (2001b) Economic impacts of transboundary plant pests and animal diseases, in *The State of Food and Agriculture 2001.* Food and Agricultural Organization of the UN, Rome, Part 111, pp. 199–226.

Farmer, A. (1997) *Managing Environmental Pollution.* Routledge, London.

Faruqui, N.I., Biswas, A.K. and Bino, M.J. (eds) (2001) *Water Management in Islam.* United Nations University Press, Tokyo.

Farvar, M.T. and Milton, J.P. (eds) (1972) *The Careless Technology.* Garden City Press, New York (NY).

Febvre, L. (1924) *A Geographical Introduction to History.* Routledge and Kegan Paul, London.

Fearnside, P. (2001) Soybean cultivation as a threat to the environment in Brazil. *Environmental Conservation* 28(1), 23–18.

Firkis, V. (1993) *Nature, Technology and Society: cultural roots of the current environmental crisis*. Adamantine Press, London.

Fischer, T.R. and Seaton, K. (2002) Strategic environmental assessment: effective planning instrument or lost concept? *Planning Practice & Research* 17(1), 31–44.

Foster, P. (1992) *The World Food Problem: tackling the causes of undernourishment in the Third World*. Lynne Rienner, Boulder (CO).

Francis, D.G. (1994) *Family Agriculture: tradition and transformation*. Earthscan, London.

Frankel, O.H., Brown, A.H.D. and Burdon, J.J. (1995) *The Conservation of Plant Biodiversity*. Cambridge University Press, Cambridge.

Freeman, C. and Jahoda, M. (eds) (1978) *World Futures: the great debate*. Martin Robertson, Oxford.

French, H. (2000) *Vanishing Borders: protecting the planet in an age of globalization*. Earthscan, London.

Fricker, A. (2000) Rejoinder: sustainable agriculture. *Futures* 32(2000), 941–2.

Friedman, J. and Weaver, C. (1979) *Territory and Function: the evolution of regional planning*. Arnold, London.

Friend, A.M. (1996) Sustainable development indicators: exploring the objective function. *Chemosphere* 33(9), 1865–87.

Foley, G. (1988) Deep ecology and subjectivity. *The Ecologist* 18(4/5), 120–3.

Fox, M. (1983) *Original Blessing*. Bear and Co., Santa Fe (NM).

Frosch, R.A. (1995) The Industrial ecology of the 21st century. *Scientific American* Sept 1995, 178–81.

Fuller, T. (2000) Will small become beautiful? *Futures* 32 (2000), 79–89.

Funtowicz, S.O. and Ravetz, J.R. (1992) The good, the true and the post-modern. *Futures* 24 (1992), 963–76.

Furguson, A.R.B. (1999) The logical foundations of ecological footprints. *Environment, Development and Sustainability* 1(2), 149–56.

Gadgil, M. and Guha, R. (1992) *This Fissured Land: an ecological history of India*. Oxford University Press, New Delhi.

Gadgil, M. and Guha, R. (1995) *Ecology and Equity: the use and abuse of nature in India*. Routledge, London.

Gavine, F.M., Rennis, D.S. and Windmill, D. (1996) Implementing environmental management systems in the finfish aquaculture industry. *Journal of the Chartered Institution of Water and Environmental Management* 10(5), 341–7.

GEMI (1998) *Environment: value to business*. Global Environmental Management Institute, Washington (DC).

Georgiou, S., Whittington, D., Pearce, D. and Moran, D. (eds) (1997) *Economic Values and the Environment in the Developing World*. Edward Elgar, Cheltenham.

Ghai, D. and Vivian, J.M. (eds) (1992) *Grassroots Environmental Action: people's participation in sustainable development*. London, Routledge.

Glantz, M.H. (ed.) (1994) *Drought Follows the Plough: cultivating marginal areas*. Cambridge University Press, Cambridge.

Glaeser, B. (ed.) (1984) *Ecodevelopment: concepts, policies, strategies*. Pergamon, New York (NY).

Glaeser, B. (1995) *Environment, Development, Agriculture*. University College London Press, London.

Glantz, M. (ed.) (2002) *La Niña and its Impacts: facts and speculation*. United Nations University Press, Tokyo.

Gleick, P.H. (ed.) (1993) *Water in Crisis: guide to the world's water resources*. Oxford University Press, Oxford.

Goldsmith, E., Allan, R., Allaby, M., Davol, J. and Lawrence, S. (1972) *Blueprint for Survival*. Harmondsworth, Penguin (also published in *The Ecologist* 2(1), 1–43).

Goldsmith, E. (1988) The way: an ecological worldview. *The Ecologist* 18 (4/5), 160–67.

Goldsmith, E. (1990) Evolution, neo-Darwinism and the paradigm of science. *The Ecologist* 20(2), 67–73.

Goodland, R. (1997) The strategic environmental assessment family. *Environmental Assessment* 5(3), 17–20.

Goodland, R. (2000) *Social and Environmental Assessment to Promote Sustainability: an informal view from the World Bank* (Environment Department Papers No. 74). World Bank, Washington (DC).

Goodland, R., Watson, C. and Ledec, G. (1984) *Environmental Management in Tropical Agriculture*. Bowker, Epping.

Goodman, D. and Redclift, M. (1991) *Refashioning Nature: food, ecology and culture*. Routledge, London.

Goudie, A. (1989) The changing human impact, in L. Friday and R. Luskey (eds) *The Fragile Environment: the Darwin College Lectures*. Cambridge University Press, Cambridge, pp. 1–20.

Gould, S.J. (1990) *Wonderful Life: the Burgess Shale and the nature of history*. Penguin, London.

Graedel, T.E. and Allenby, B.R. (eds) (1995) *Industrial Ecology*. Prentice Hall, Englewood Cliffs (NJ).

Graham-Rowe, D. (2003) Bug dines on dioxins. *New Scientist* 177(2379), 20.

Gray, R.H., Owen, D.L. and Adams, C. (eds) (1996) *Accounting and Accountability: social and environmental accounting in a changing world*. Prentice-Hall, Hemel Hempstead.

Grigg, D. (1985) *The World Food Problem*. Blackwell, Oxford.

Groombridge, B. and Jenkins, M. (2002) *World Atlas of Biodiversity*. University of California Press, Berkeley (CA).

Grove, R.H. (1992) The origins of western environmentalism. *Scientific American* 267(1), 22–7.

Grove, R.H. (1995) *Green Imperialism: colonial expansion, tropical island Edens and the origins of environmentalism, 1600–1860*. Cambridge University Press, Cambridge.

Gupta, A. and Asher, M.G. (1998) *Environment and the Developing World: principles, policies and management*. Wiley, Chichester.

Gupta. A. (1988) *Ecology and Development*. Routledge, London.

Gupta, J. (2001) *Our Simmering Planet: what to do about global warming?* Zed, London.

Guthman, J. (1997) 'Representing Crisis': the theory of Himalayan environmental degradation and the Project of Development in Post-Rana Nepal. *Development & Change* 28(1), 45–69.

Haab, T.C. and McConnel, K.E. (2002) *Valuing Environmental and Natural Resources: the economics of non-market valuation*. Edward Elgar, Cheltenham.

Hanley, N., Moffatt, I., Faichney, R. and Wilson, M. (1999) Measuring sustainability: a time series of alternative indicators for Scotland. *Ecological Economics* 28(1), 55–73.

Hannam, K. (1999) Environmental management in India: recent challenges to the Indian Forest Service. *Journal of Environmental Planning and Management* 42, 221–33.

Hardin, G. (1968) The tragedy of the commons. *Science* 162(3859), 1243–8.

Hardin, G. (1974a) Lifeboat ethics: the case against helping the poor. *Psychology Today* 8, 38–43, 123–6.

Hardin, G. (1974b) *The Ethics of a Lifeboat*. American Association for the Advancement of Science, Washington (DC).

Hardoy, J.E., Mitlin, D. and Satterthwaite, D. (1992) *Environmental Problems in Third World Cities*. Earthscan, London.

Hardoy, J.E. and Satterthwaite, D. (1989) *Squatter Citizen: life in the urban Third World*. Earthscan, London.

Hardoy, J.E., Satterthwaite, D. and Cairncross, S. (eds) (1990) *The Poor Die Young: housing and health in Third World cities*. Earthscan, London.

Harremoës, P., Gee, D., MacGarvin, M., Stirling, A., Keys, J., Wynne, B. and Guedes Vaz, S. (2002) *The Precautionary Principle in the 20th Century: late lessons from early warnings*. Earthscan, London.

Harris, N. (1990) *Environmental Issues in the Cities of the Developing World*. DPU Working Papers No. 20. Development Planning Unit, University College London, London.

Harrison, P. (1993) *The Third Revolution: environment, population and a sustainable world*. Penguin, Harmondsworth.

Harvey, B. and Hallett, J.D. (1977) *Environment and Society: an introductory analysis*. Macmillan, London.

Harvey, D. (1989) *The Condition of Postmodernity: an enquiry into the origins of cultural change*. Basil Blackwell, Oxford.

Hasan, S. and Khan, M.A. (1999) Community-based environmental management in a megacity: considering Calcutta. *Cities* 16(2), 103–10.

Hasler, R. (1999) *An Overview of the Social, Ecological and Economic Achievements of Zimbabwe's CAMPFIRE Programme*. IIED 'Evaluating Eden' Discussion Papers No. 3. International Institute for Environment and Development, London.

Heiskanen, E. (2002) The institutional logic of life cycle assessment. *Journal of Cleaner Production* 10, 427–37.

Helm, D. (ed.) (1991) *Economic Policy towards the Environment*. Blackwell, Oxford.

Hettne, B. (1990) *Development and the Three Worlds*. Longman, Harlow.

Hewitt, K. (ed.) (1983) *Interpretations of Calamity: from the viewpoint of human ecology*. Allen and Unwin, Boston (MA).

Hewitt, K. (1997) *Regions at Risk: a geographical introduction to disasters*. Longman, Harlow.

Heyward, V. (1995) *Global Biodiversity Assessment*. Cambridge University Press, Cambridge.

Hill, J.C. (1964) Puritanism, capitalism and the scientific revolution. *Past and Present* 29, 88–97.

Holdgate, M. (1979) *A Perspective of Environmental Pollution*. Cambridge University Press, Cambridge.

Holdgate, M. (1999) *The Green Web: a union for world conservation*. Earthscan, London.

Holling, C.S. (1978) *Adaptive Environmental Assessment and Management*. Wiley, New York (NY).

Holmberg, J., Thomson, K. and Timberlake, L. (1993) *Facing the Future: beyond the Earth Summit*. London, Earthscan.

Honkasalo, A. (1998) The EMAS scheme: a management tool and instrument of environmental policy. *Journal of Cleaner Production* 6, 119–28.

Honigsbaum, M. (2001) *The Fever Trail: the hunt for the cure for malaria*. Macmillan, London.

Hughes, B.B. (1980) *World Modelling: the Mesarovic–Pestel world model in the context of its contemporaries*. Lexington Books, Lexington (MS).

Hunt, D. and Johnson, C. (1995) *Environmental Management Systems: principles and practice*. McGraw-Hill, New York (NY).

Hunt, E. (2004) *Thirsty Planet: strategies for sustainable water management*. Zed, London.

Huntington, E. (1915) *Civilisation and Climate*. Yale University Press, New Haven (CO).

Hurley, A. (1995) *Environmental Inequalities: class, race, and industrial pollution in Gary, Indiana, 1945–1980*. University of North Carolina Press, Chapel Hill (NC).

Hutton, J. and Dickson, B. (eds) (2000) *Endangered Species, Threatened Convention. The past, present and future of CITES*. Earthscan, London.

Ince, M. (1991) *The Rising Seas*. Earthscan, London.

Independent Commission on International Development Issues (1980) *North–South: a programme for survival*. Pan, London.

IUCN (1997) *Indigenous Peoples and Sustainability: cases and actions* (IUCN Inter-Commission Task Force on Indigenous Peoples). International Books, Utrecht.

IUCN, UNEP and WWF (1980) *World Conservation Strategy: living resources for sustainable development*. International Union for Conservation of Nature and Natural Resources, Gland.

Ives, J.D. (1987) The theory of Himalayan environmental degradation: its validity and application challenged by recent research. *Mountain Research and Development* 7, 185–99.

Ives, J.D. and Messerlie, B. (1989a) *The Theory of Himalayan Environmental Degradation. What is the nature of the perceived crisis?* Routledge, New York (NY).

Ives, J.D. and Messerlie, B. (1989b) *The Himalayan Dilemma: reconciling development and conservation*. Routledge, New York (NY).

Ives, J.H. (ed.) (1985) *The Export of Hazards: transnational corporations and environmental control*. Routledge & Kegan Paul, Boston (MA).

Jacobs, M. (1991) *The Green Economy*. Pluto Press, London.

Jacobson, J.K. (1992) *Gender Bias: roadblock to sustainable development*. The Worldwatch Institute, Washington (DC).

Jansen, K. and Vellema, S. (eds) (2004) *Agribusiness and Society: corporate responses to environmentalism, market opportunities and public regulation*. Zed, London.

Jasch, C. (2000) Environmental performance evaluation and indicators. *Journal of Cleaner Production* 8(1), 79–88.

Jasanoff, S. (ed.) (1994) *Learning from Disaster: risk management after Bhopal*. University of Pennsylvania Press, Philadelphia (PA).

Jeffrey, D.W. and Madden, B. (eds) (1991) *Bioindicators and Environmental Management*. Academic Press, London.

Jeffery, R. and Vira, B. (eds) (2001) *Conflict and Cooperation in Participatory Natural Resource Management*. Palgrave, NewYork (NY).

Jiliberto, R. (2002) Decisional environmental values as the object of analysis for strategic environmental assessment. *Impact Assessment and Project Appraisal* 20(1), 61–70.

Johnston, R. (1989) *Environmental Problems: native, economy and state*. Belhaven, London.

Kadomura, H. (1997) *Data Book of Desertification/Land Degradation*. Centre for Global Environmental Research, Tokyo.

Kahn, H., Brown, W. and Martel, L. (eds) (1976) *The Next 200 Years*. Abacus, London.

Kapoor, I. (2001) Towards participatory environmental management? *Journal of Environmental Management* 63(3), 269–79.

Karagozoglu, N. and Lindell, M. (2000) Environmental management: testing the win–win model. *Journal of Environmental Planning and Management* 43(6), 817–30.

Kasperson, J.X., Kasperson, R.E. and Turner II, B.L. (eds) (2003) *Regions at Risk: comparisons of threatened environments*. United Nations University Press, Tokyo.

Kassas, M. (1999) Rescuing drylands: a project for the world. *Futures* 31 (1999), 945–58.

Kassas, M. (2001) *The Water Crisis*. Kluwer, Dordrecht.

Kate, K.-Ten and Laird, S.A. (2002) *The Commercial Use of Biodiversity: access to genetic resources and benefit sharing*. Earthscan, London.

Kates, R. and Clark, W. (1996) Environmental surprise: expecting the unexpected. *Environment* 38, 6–11, 28–34.

Keeney, R.L. and von Winterfeldt, D. (2001) Appraising the precautionary principle: a decision analysis perspective. *Journal of Risk Research* 4(2), 191–202.

Keys, D. (1999) *Catastrophes: an investigation into the origins of the modern world*. Century Books, London.

Keil, R., Bell, D.V.J., Penz, P. and Fawcett, L. (eds) (1998) *Political Ecology: global and local*. Routledge, London.

Khoo, C.H. (1991) Environmental management in Singapore. *Environmental Monitoring and Assessment* 19(1–3), 127–30.

Khor, M. (2004) *Intellectual Property, Biodiversity and Sustainable Development: resolving the difficulties*. Zed, London.

Killham, K. (1994) *Soil Ecology*. Cambridge University Press, Cambridge.

Kirkpatrick, C., Clarke, R. and Polidano, C. (eds) (2002) *Handbook on Development Policy and Management*. Edward Elgar, Cheltenham.

Kirkpatrick, C. and Lee, N. (eds) (1993) *Sustainable Development in a Developing World: integrating socio-economic and environmental management*. Edward Elgar, Cheltenham.

Kirkpatrick, D. (1990) Environmentalism: the new crusade. *Fortune* February 12: 24–30.

Klare, M.T. (2001) *Resource Wars: the new landscape of global conflict*. Henry Holt, New York (NY).

Koziel, I. and Saunders, J. (eds) (2001) *Living off Biodiversity: exploring livelihoods and biodiversity issues in natural resource management*. IIED, London.

Kriebel, D., Tickner, J., Epstein, P., Lemons, J., Levins, R., Loechler, E., Quinn, M., Rudel, R., Schettler, T. and Stoto, M. (2001) The precautionary principle in environmental science. *Environmental Health Perspectives* 109(9), 871–6.

Kuzmiak, D.T. (1991) A history of the American environmental movement. *The Geographical Journal* 157(3): 265–78.

Kropotkin, P. (1974) *Fields, Factories and Workshops*. Unwin, London (edited by C. Ward; original *Fields, Factories and Workshops Tomorrow*, 1899, in Russian).

Le Roi Ladurie, E. (1972) *Times of Feast, Times of Famine: a history of climate since the year 1000*. (trans. B. Brag). George Allen & Unwin, London (original French edn 1967 *Histoire du climat depuis l'an mil*. Flammarion, Paris).

Lafferty, W.M. and Meadowcroft, J. (eds) (1996) *Democracy and the Environment: problems and prospects*. Edward Elgar, Cheltenham.

Lapierre, D. and Moro, D. (2002) *Five Past Midnight in Bhopal*. (trans. K. Spink) Scribner, London (original French edn 2001, Pressinter SA and Sesamat SA, Paris).

Lange, G-M., Hassan, R. and Alfieri, A. (2003a) Using environmental accounts to promote sustainable development: experience in southern Africa. *Natural Resources Forum* 27(1), 19–31.

Lange, G-M., Hassan, R. and Hamilton, K. (eds) (2003b) *Environmental Accounting in Action: case studies from Southern Africa*. Edward Elgar, Cheltenham.

Latour, B. (1993) *We Have Never Been Modern*. (Trans. C. Porter)Harvester Wheatsheaf, Hemel Hempstead .

Laurance, W.F., Albernaz, A.K.M. and Da Costa, C. (2001) Is deforestation accelerating in the Brazilian Amazon? *Environmental Conservation* 28(4), 305–11.

Lave, L.B. (ed.) (1987) *Risk Assessment and Management*. Plenum Press, New York (NY).

Leach, G. and Mearns, R. (1989) *Beyond the Woodfuel Crisis: people, land and trees in Africa*. Earthscan, London.

Leach, M. and Mearns, R. (1996) *The Lie of the Land: challenging received wisdom on the African continent.* James Currey for The International African Institute, Oxford, London.

Leach, M., Mearns, R. and Scoones, I. (eds) (1997) Community-based sustainable development: consensus or conflict? *IDS Bulletin* 28(4), 1–96.

Lee, N. and Kirkpatrick, C. (eds) (2000) Integrated Appraisal and Sustainable Development in a Developing World. Edward Elgar, Cheltenham.

Lee, N. and Walsh, F. (1992) Strategic environmental assessment: an overview. *Project Appraisal* 7(3), 126–36.

Leitmann, J. (1996) Browning the Bank: the World Bank's growing investment in urban environmental management. *Environmental Impact Assessment Review* 16(4–6), 351–61.

Lele, S. (1991) Sustainable Development: a critical review. *World Development* 19(6), 607–21.

Leopold, A. (1949) *A Sand County Almanac.* London, Oxford University Press.

Lewin, R. (1993) *Complexity: life at the edge of chaos.* London, Dent.

Lewis, C. (ed.) (1996) *Managing Conflicts in Protected Areas.* IUCN, Gland.

Lewis, M. (1992) *Green Delusions: an environmentalist critique of radical environmentalism.* Duke University Press, Durham (NC).

Lipton, M. (1977) *Why Poor people Stay Poor: urban bias in world development.* Temple Smith, London.

Lipton, M. and Longhurst, R. (1989) *New Seeds and Poor People.* Unwin-Hyman, London.

Lomborg, B. (2001) *The Sceptical Environmentalist: measuring the real state of the world.* (trans. H. Mathews) Cambridge University Press, Cambridge (Original Danish edn 1998).

Lopez, P. (1999) *Incorporating Developing Countries into Global Efforts for Greenhouse Gas Reduction.* Resources for the Future Climate Issue Brief No. 16. Resources for the Future, Washington (DC).

Low, P. (ed.) (1992) *International Trade and the Environment* (World Bank Discussion Paper).World Bank, Washington (DC).

Lowe, E.A., Warren, J.L. and Moran, S.R. (2000) *Discovering Industrial Ecology: an executive briefing and sourcebook.* Battelle Press, Columbus (OH).

Lowe, J. and Lewis, D. (1980) *The Economics of Environmental Management.* Philip Alan, Oxford.

Lundqvist, J. (2000) A global perspective on water and the environment. *Physical Chemistry of the Earth (B)* 2(3), 259–64.

Mabogunje, A.L. (1980) *The Development Process: a spatial perspective.* Hutchinson, London.

Mackenzie, D. (2002) Fresh evidence on Bhopal disaster. *New Scientist* 176(2372), 6–7.

Mackenzie, F. (1993) Exploring the connections: structural adjustment, gender and the environment. *Geoforum* 24(1), 71–87.

Maddox, J. (1972) *The Doomsday Syndrome.* Maddox Educational Ltd., London.

Malmqvist, B. and Rundle, S. (2002) Threats to the running water ecosystems of the world. *Environmental Conservation* 29(2), 134–53.

Mannion, A.M. (1991) *Global Environmental Change: a natural and cultural environmental history.* Longman, Harlow.

Mara, D. and Cairncross, S. (1990) *Guidelines for the Safe Use of Wastewater and Excreta in Agriculture and Aquaculture.* World Health Organisation, Geneva.

Marsh, G.P. (1864) *Man and Nature: or physical geography as modified by human action.* Charles Scribner, New York (NY).

Marshall, F., Ashore, M. and Hinchcliffe, F. (1997) *A Hidden Threat to Food Production: air pollution and agriculture in the developing world.* International Institute for Environment and Development (Gatekeeper Series No.73), 3 Endsleigh Street, London.

Martens, W.J.M. (1998) *Health and Climate Change: monitoring the impacts of global warming and ozone depletion*. Earthscan, London.

Martin, P.S. (1967) Prehistoric overkill, in P.S. Martin and H.E. Wright (eds) *Pleistocene Extinctions*. Yale University Press, New Haven (CO), pp. 75–92.

Martin, P.S. and Klein, R.G. (eds) (1984) *Quaternary Extinctions*. University of Arizona Press, Tucson (AZ).

Martin, S. (2002) Professionals and sustainability. *Journal of the Institution of Environmental Sciences* 11(3), 6–7.

Maskrey, A. (1989) *Disaster Mitigation: a community based approach*. Oxford Development Guidelines No. 3. Oxfam, Oxford.

Mason, C. (2003) *The 2030 Spike: countdown to global catastrophe*. Earthscan, London.

May, P.J., Burby, R.J., Eriksen, N.J., Handmer, J.W., Dixon, J.E., Michaels, S. and Ingle Smith, D. (1996) *Environmental Management and Governance: intergovernmental approaches to hazards and sustainability*. Routledge, London.

McCall, G.S.H., Larning, D. and Scott, S. (eds) (1991) *Geohazards*. Chapman and Hall, London.

McConnel, F. (1996) *The Biodiversity Convention: a negotiating history*. Kluwer Law International, London.

McCormick, J. (1988) *Acid Earth: the global threat of acid pollution*. Earthscan, London.

McCormick, J. (1989) *Reclaiming Paradise: the global environmental movement*. Indiana University Press, Bloomington (IN).

McCulloch, J. (ed.) (1990) *Cities and Global Climate Change*. Climate Institute, Washington (DC).

McCully, P. (2001) *Silenced Rivers: the ecology and politics of large dams* (updated edn). Zed, London.

McEvoy, III, J. (1971) A comment: conservation – an upper-middle class social movement. *Journal of Leisure Research*. 3: 127–8.

McGranaham, G. (1991) *Environmental Problems and the Urban Household in Third World Countries*. The Stockholm Environmental Institute, Stockholm.

McNeill, J. (2000) *Something New Under the Sun: an environmental history of the twentieth century*. Penguin, London.

Meadows, D.H., Meadows, D.L., Randers, J. and Behrens, III, W.W. (1972) *The Limits to Growth* (a report for the Club of Rome's project on the predicament of mankind). Universal Books, New York (NY).

Meadows, D.H., Meadows, D.L. and Randers, J. (1992) *Beyond the Limits: global collapse or sustainable future?* Earthscan, London.

Merchant, C. (1992) *Radical Ecology: the search for a liveable world*. Routledge, London.

Mermet, B.R. (2002) Integrated coastal management at the regional level: lessons from Tolia, Madagascar. *Ocean & Coastal Management* 45(1), 41–58.

Merton, R.K. (1970) *Science, Technology and Society in Seventeenth Century England*. Howard Fertig, New York (NY).

Mesarovic, M. and Pestel, E. (1975) *Mankind at the Turning Point* (the second report to the Club of Rome). Hutchinson, London (published 1974 in USA).

Messerschmidt, D.A. (1990) Indigenous environmental management and adaptation: an introduction to four case studies from Nepal. *Mountain Research and Development* 10(1), 3–4.

Milani, B. (2000) *Designing the Green Economy: the post-industrial alternative to corporate globalization*. Rowman & Littlefield, Lanham (MD).

Miller, jnr, G.T. (1990) *Living in the Environment: an introduction to environmental science* (6th edn). Wadsworth, Belmont (CA).

Miller, M.A.L. (1995) *The Third World in Global Environmental Politics.* Open University Press, Buckingham.

Mitsch, W.J. and Gosselink, J.G. (1993) *Wetlands* (2nd edn). Van Nostrand Reinhold, New York (NY).

Mohammed, S.M. (2002) Pollution management in Zanzibar: the need for a new approach. *Ocean & Coastal Management* 45(4–5), 301–11.

Mokhtar, M.B. and Ghani Aziz, S.A.B.A. (2003) Integrated coastal zone management using the ecosystem approach: some perspectives in Malaysia. *Ocean and Coastal Zone Management* 46(5), 407–19.

Moore, G. and Jennings, S. (eds) (2000) *Commercial Fishing: the wider ecological impacts.* Blackwell Science, London.

Morehouse, W. (1994) Unfinished business: Bhopal ten years after. *The Ecologist* 24(5), 164–8.

Morris, W. (1891) *News from Nowhere.* Reeves and Turner, London.

Morrison, D.E. (1986) How and why environmental consciousness has trickled down, in A. Schnaiberg, N. Watts and K. Zimmerman (eds) *Distributional Conflicts in Environmental Resource Policy.* Gower, Aldershot, pp. 187–220.

Morton, J. (2002) Drought management as growth industry and as a development paradigm. Paper presented to the Development Studies Association Annual Conference, Greenwich University, London, 9 November 2002 (mimeo 18 pp.).

Mowforth, M. and Munt, I. (2003) *Tourism and Sustainability: new tourism in the Third World* (2nd edn). Routledge, London.

Moxen, J. and Strachan, P. (1995) The formulation of standards for environmental management systems: structural and cultural issues. *Greener Management International* 12(October), 32–48.

Munasinghe, M. and McNeely, J.A. (eds) (1994) *Protected Area Economics and Policy: linking conservation and sustainable development.* World Bank, Washington (DC).

Munn, R.E. (1979) *Environmental Impact Assessment* (SCOPE 5) (2nd edn). Wiley, Chichester.

Munn, R.E. (2002) Cumulative environmental assessment, in R.E. Munn (ed.) *Encyclopedia of Global Environmental Change*, vol. 4. Wiley, Chichester, pp. 178–80.

Myers, N. (1985) *The Primary Source: tropical forests and our future.* W.W. Norton, New York (NY).

Myers, N. (2004) *Banana Wars: The Price of Free Trade.* Zed, London.

Naess, A. (1973) The shallow and the deep: long range ecology movement – a summary. *Inquiry* 16, 95–100.

Naess, A. (1988) Deep ecology and ultimate premises. *The Ecologist* 18(4/5), 128–32.

Naess, A. (1989) *Ecology, Community and Lifestyles: outline of an ecosophy.* Cambridge University Press, Cambridge.

New Scientist (2003). In deep trouble (editorial). *New Scientist* 178(2395), 3.

Newby, H. (1990) Environmental change and the social sciences. Paper to 1990 Annual Meeting of the British Association for the Advancement of Science (mimeo., 22 pp).

Newson, M. (1992) *Land, Water and Development: river basin systems and their sustainable management.* Routledge, London.

Nicholson, E.M. (1970) *The Environmental Revolution.* Hodder and Stoughton, London.

Niemeijer, D. and Mazzucato, V. (2002) Soil degradation in the West African Sahel: how serious is it? *Environment* 44(2), 20–31.

Nijkamp, P. and Soeteman, F. (1998) Ecologically sustainable economic development: key issues for strategic environmental management. *International Journal of Social Economics* 15(3–4), 88–102.

Nitz, T. and Holland, I. (2000) Does environmental impact assessment facilitate environmental management activities? *Journal of Environmental Assessment Policy and Management* 2(1), 1–17.

Noble, B.F. (2000) Strategic environmental assessment: what is it and what makes it strategic? *Journal of Environmental Assessment Policy and Management* 2(2), 203–24.

Norton, B.G. and Steinemann, A.C. (2001) Environmental values and adaptive management. *Environmental Values* 10(4), 437–506.

Norton, T.W., Beer, T. and Dovers, S.R. (eds) (1996) *Risk and Uncertainty in Environmental Management* (Proceedings of the 1995 Australian Academy of Science, Fenner Conference on the Environment). Centre for Resource and Environmental Studies, Australian National University, Canberra.

Oates, J.F. (1995) The dangers of conservation by rural development: a case study from the forests of Nigeria. *Oryx* 229(2), 115–22.

Oates, J.F. (1999) *Myth and Reality in the Rain Forest: how conservation strategies are failing in West Africa.* University of California Press, Berkeley (CA).

Oberthür, S., Ott, S.E. and Tarasofsky, R.G. (1999) *The Kyoto Protocol: international climate policy for the 21st century.* Springer, Berlin.

Ohlsson, L. (ed.) (1995) *Hydropolitics: conflicts over water as a development constraint.* Zed, London.

Oldfield, S. (1988) *Buffer Zone Management in Tropical Moist Forest.* IUCN, Gland.

Oliver-Smith, A. (ed.) (1986) *Natural Disasters and Cultural Responses.* Studies in Third World Societies No. 36, Department of Anthropology, College of William and Mary, Williamsburg (VA).

O'Riordan, T. (1976) *Environmentalism.* Pion, London.

O'Riordan, T. (1995) *Environmental Science for Environmental Management.* Longman, Harlow.

O'Riordan, T. and Cameron, J. (1994) *The History and Contemporary Significance of the Precautionary Principle.* Earthscan, London.

O'Riordan, T. and Stoll-Kleemann, S. (eds) (2001) *Biodiversity, Sustainability and Human Communities.* Cambridge University Press, Cambridge.

Orstrom, E. (1990) *Governing the Commons: the evolution of institutions for collective action.* Cambridge University Press, New York (NY).

Osborn, F. (1948) *Our Plundered Planet.* Little Brown and Co., Boston (MD).

Osborn, F. (1953) *Limits of the Earth.* Little Brown and Co., Boston (MD).

Osuagwu, L. (2002) TQM strategies in a developing economy: empirical evidence from Nigerian companies. *Business Process Management Journal* 8(2), 160–81.

Oudshoorn, H.M. (1997) The pending water crisis. *GeoJournal* 42(1), 27–38.

Palmer, A.R. (1999) Ecological Footprints: evaluating sustainability. *Environmental Geosciences* 6(4), 200–4.

Park, C.C. (1980) *Ecology and Environmental Management.* Dawson and Sons, Folkstone.

Park, C.C. (1987) *Acid Rain: rhetoric and reality.* Methuen, London.

Park, M. (1969) *Travelling in Africa.* Dent, London (originally published 1811).

Partidário, M. and Clark, R. (eds) (2000) *Perspectives on Strategic Environmental Assessment.* Lewis Publishers, Boca Raton (FL).

Passmore, J. (1974) *Man's Responsibility for Nature.* Duckworth, London.

Patterson, M. (1996) *Global Warming and Global Politics.* Routledge, London.

Pearce, D.W. and Barbier, E.B. (2002) *Blueprints for a Sustainable Economy.* Earthscan, London.

Pearce, D.W., Barbier, E.B., Markandya, A., Barrett, S., Turner, R.K. and Swanson, T. (1991) *Blueprint 2: greening the world economy.* Earthscan, London.

Pearce, D.W., Barbier, E.B., Markandya, A., Barrett, S., Turner, R.K. and Swanson, T. (1993) *Blueprint 3: measuring sustainable development*. Earthscan, London.

Pearce, D.W., Markandya, A. and Barbier, E.B. (1989) *Blueprint for a Green Economy*. Earthscan, London.

Pearce, D.W. and Turner, R.K. (1990) *Economics of Natural Resources and the Environment*. Harvester Wheatsheaf, Hemel Hempstead.

Pearce, F. (2002a) Africans go back to the land as plants reclaim the desert. *New Scientist* 175(2361), 4–5.

Pearce, F. (2002b) Pollution is plunging us into darkness. *New Scientist*, 176(2373), 6–7.

Pearce, F. (2003) Dismay over call to build new dams. *New Scientist* 177(2387), 11.

Pearce, F. (2004) Gene pollution is 'pervasive'. *New Scientist* 181[2436], 8.

Peat, D. (1988) *Superstrings and the Search for the Theory of Everything*. Cardinal-Sphere Books, London.

Peet, R. and Watts, M. (2004) *Liberation Ecologies* (2nd edn). Routledge, London.

Pentreath, R.J. (2000) Strategic environmental management: time for a new approach. *The Science of the Total Environment* 249(1), 3–11.

Pepper, D. (1984) *The Roots of Modern Environmentalism*. Croom Helm, London.

Perera, L. and Amin, A. (1996) Accommodating the informal sector: a strategy for urban environmental management. *Journal of Environmental Management* 46(1), 3–15.

Perrings, C., Mahler, K.G., Folke, C., Holling, C.S. and Jansson, B.O. (eds) (1995) *Biodiversity Loss: economic and ecological issues*. Cambridge University Press, Cambridge.

Perrings, C. and Vincent, J.R. (eds) (2004) *Natural Resource Accounting and Economic Development: theory and practice*. Edward Elgar, Cheltenham.

Pest, C. and Grabber, J. (2001) Historical analysis, a valuable tool in community-based environmental protection. *Marine Pollution Bulletin* 42(5), 339–49.

Philander, S.G. (1998) *Is the Temperature Rising? The uncertain science of global warming*. Princeton University Press, Princeton (NJ).

Phillimore, J. and Davidson, A. (2002) A cautionary tale: Y2K and the politics of foresight. *Futures* 34 (2002), 147–57.

Pierce, J.T. (1990) *The Food Resource*. Longman, Harlow.

Pigou, A.C. (1932) *The Economics of Welfare* (first edn. 1920). Macmillan, London.

Pimbert, M. and Pretty, J. (1995) *Parks, People and Professionals: putting 'participation' into protected area management*. United Nations Research Institute for Social Development Discussion Paper No. 57. UNRISD, Geneva.

Pinchot, G. (1910) *The Fight for Conservation*. University of Washington Press, Seattle (WA).

Polasky, S. (ed.) (2002) *Economics of Biodiversity Conservation*. Ashgate, Aldershot.

Ponting, C. (1991) *A Green History of the World*. Sinclair-Stevenson, London.

Porter, A.L. (1995) Technology Assessment. *Impact Assessment* 13(2), 135–51.

Porter, R.N. and Brownlie, S.F. (1990) Integrated environmental management: a planning strategy for nature conservation developments. *South African Journal of Wildlife Research* 20(2), 81–6.

Posey, D.A. (1990) Intellectual Property Rights and Just Compensation for Indigenous Knowledge: challenges to science and international law. Paper to the Association for Applied Anthropology, New York (NY).

Posey, D.A. (1999) *Cultural and Spiritual Values of Biodiversity* (published for the UNEP). Intermediate Technology Publications, London.

Postel, S. (1992) *Last Oasis, Facing Water Scarcity*. W.W. Norton, New York (NY).

Power, A.G. (1999) Linking ecological sustainability and world food needs. *Environment, Development and Sustainability* 1(1), 185–96.

Pretty, J.N. and Howes. R. (1993) *Sustainable Agriculture in Britain: recent advances and new policy challenges.* IIED Research Series 2(1), International Institute for Environment and Development, London.

Pretty, J.N. (1999) Can sustainable agriculture feed Africa? New evidence on progress, processes and impacts. *Environment, Development and Sustainability* 1(1), 253–74.

Pretty, J.N. and Ward, H. (2001) Social capital and the environment. *World Development* 29(2), 209–27.

Princen, T. and Finger, M. (1995) *Environmental NGOs in World Politics: linking the local and the global.* Routledge, London.

Progress Publishers (1977) *Current Problems, Society and Environment: a Soviet view* (trans. J. Williams). Progress Publishers, Moscow.

Pryde, P.R. (1991) *Environmental Management in the Soviet Union.* Cambridge University Press, Cambridge.

Pye-Smith, C. (2003) Fruits of the forest. *New Scientist* 179(2404), 36–9.

Ravetz, J. (2000a) How I got Y2K wrong. *Futures* 32 (2000), 937–9.

Ravetz, J. (2000b) Integrated assessment for sustainability appraisal in cities and regions. *Environmental Impact Assessment Review* 20, 31–64.

Raymond, K. (1996) The long-term future of the Great Barrier Reef. *Futures* 28(1996), 947–70.

Redclift, M. (1984) *Development and the Environmental Crisis.* Methuen, London.

Redclift, M. (1995) The environment and structural adjustment: lessons for policy intervention in the 1990s. *Journal of Environmental Management* 44(1), 55–68.

Reed, D. (ed.) (1992) *Structural Adjustment and the Environment.* Earthscan, London.

Reed, D. (ed.) (1996) *Structural Adjustment, the Environment, and Sustainable Development.* Earthscan, London.

Rees, J. (1985) *Natural Resources: allocation, economics and policy.* London, Methuen.

Rees, W.E. (1992) Ecological footprints and appropriate carrying capacity: what urban economics leaves out. *Environment and Urbanization* 41(2), 121–30.

Reij, C., Scoones, I. and Toulmin, C. (eds) (1996) *Sustaining the Soil: indigenous soil and water conservation in Africa.* Earthscan, London.

Reijntjes, C., Haverkort, B. and Waters-Bayer, A. (1992) *Farming for the Future: an introduction to low-external input and sustainable agriculture.* Macmillan, London.

Reith, C. (2001) Applying environmental management strategies to the agricultural sector: Louisiana's model sustainable agricultural complex. *Corporate Environmental Strategy* 8(1), 75–83.

Riddell, R. (1981) *Ecodevelopment: economics, ecology and development.* Gower, Farnborough.

Rietbergen-McCraken, J. and Abaza, H. (eds) (2000) *Environmental Valuation: a worldwide compendium of case studies.* Earthscan, London.

Rigby, D., Woodhouse, P., Young, T. and Burton, M. (2001) Constructing a farm level indicator of sustainable agricultural practice. *Ecological Economics* 39(4), 463–78.

Robbins, P. (2004) *Stolen Fruit: the tropical commodities disaster.* Zed, London.

Robért, K.-H. (2000) Tools and concepts for sustainable development: how do they relate to a general framework for sustainable development, and to each other? *Journal of Cleaner Production* 8, 243–54.

Robért, K.-H., Schmidt-Bleek, B., Aloisi de Lardarel, J., Basile, G., Jansen, J.L., Kuehr, R., Price Thomas, P., Suzuki, M., Hawken, P. and Wackenaegel, M. (2002) Strategic sustainable development: selection, design and synergies of applied tools. *Journal of Cleaner Production* 10, 197–214.

Robinson, D. and Kellow, A. (eds) (2001) *Globalization and the Environment: risk assessment and the WTO.* Edward Elgar, Cheltenham.

Rodhe, H. (1989) Acidification in a global perspective. *Ambio* XVIII (3), 155–60.

Rodhe, H., Callaway, J. and Dianwu, Z. (1992) Acidification in Southwest Asia: prospects for coming decades. *Ambio* XXI (2), 148–50.

Rodhe, H. and Herrera, R. (eds) (1988) *Acidification in Tropical Countries* (SCOPE No. 36). Chichester, Wiley.

Roe, E.M. (1995) Except-Africa: postscript to a special section on development narratives. *World Development* 23(6), 1065–9.

Roggeri, H. (1995) *Tropical Freshwater Wetlands: a guide to current knowledge and sustainable management.* Kluwer Academic, Dordrecht.

Rojsek, I. (2001) From red to green: towards environmental management in the countries in transition. *Journal of Business Ethics* 33(1), 37–50.

Rondinelli, D. and Vastag, G. (2000) Panacea, common sense, or just a label? The value of ISO 14001 environmental management systems. *European Management Journal*, 18(5), 499–510.

Rostow, W.W. (1960) *The Stages of Economic Growth: a non-communist manifesto.* Cambridge University Press, Cambridge.

Roth, E., Rosenthal, H. and Burbridge, P.R. (2000) A discussion of the use of the sustainability index: 'ecological footprint' for aquaculture production. *Aquatic and Living Resources* 13, 461–9.

Royal Tropical Institute (1990) *Environmental Management in the Tropics: an annotated bibliography 1985–1989.* Koninklijk Instituut voor de Tropen, Amsterdam.

Russel, C.S. (ed.) (2003) *Economics of Environmental Monitoring and Enforcement.* Ashgate, Aldershot.

Sachs, W. (ed.) (1992) *The Development Dictionary: a guide to knowledge as power.* Zed, London.

Sagar, A.M. and Najam, A. (1998) The human development index: a critical review. *Ecological Economics* 25(3), 249–64.

Said, E. (1994) *Culture and Imperialism.* Vantage, London.

Sale, K. (1985) *Dwellers in the Land: the bioregional vision.* Sierra Club, San Francisco (CA).

Sánchez, L.E. and Hacking, T. (2002) Integrative management: an approach to linking impact assessment and environmental management systems. *Impact Assessment and Project Appraisal* 20(1), 25–38.

Sandbach, F. (1984) *Environment, Ideology and Policy.* Basil Blackwell, Oxford.

Sarkar, S. (1999) *Eco-socialism or Eco-capitalism: a critical analysis of humanity's fundamental choices.* Zed, London.

Savenjie, H. and Huijsman, A. (eds) (1991) *Making Haste Slowly: strengthening local environmental management in agricultural development.* Koninklinjk Instituut voor de Tropen, The Hague.

Save the Children (1995) *Toolkits: a practical guide to assessment, monitoring, review and evaluation* (compiled by L. Gosling and M. Edwards) (revised edn published 2004). Save the Children, London.

Sawhney, A. (2004) *The New Face of Environmental Management in India.* Ashgate, Aldershot.

Schell. L.M., Smith, M.T. and Bilsborough, A. (eds) (1993) *Urban Ecology and Health in the Third World.* Cambridge University Press, New York (NY).

Schoffmann, A. and Tordini, A.M. (2000) *ISO 14001: a practical approach*. Oxford University Press, Oxford.

Schramm, G. and Warford, J.J. (eds) (1989) *Environmental Management and Economic Development*. Johns Hopkins University Press, Baltimore (MD).

Schumacher, E.F. (1973) *Small is Beautiful: a study of economics as if people mattered*. Bland and Briggs, London.

Scolimowski, H. (1988) Eco-philosophy and deep ecology. *The Ecologist* 18(4/5), 124–7.

Scoones, I. (ed.) (2001) *Dynamics & Diversity: soil fertility and family livelihoods in Africa*. Earthscan, London.

Semple, E.C. (1911) *Influences of Geographic Environment*. Henry Holt, New York (NY).

Sen, A.K. (1981) *Poverty and Famines: an essay on entitlement and deprivation*. Clarendon Press, Oxford.

Shepherd, A. (1998) *Sustainable Rural Development*. Macmillan, London.

Shepherd, R. (1980) *Prehistoric Mining and Allied Industries*. Academic Press, London.

Shi, C., Hutchinson, S.M., Yu, L. and Xu, S. (2001) Towards a sustainable coast: an integrated coastal zone management framework for Shanghai, People's Republic of China. *Ocean & Coastal Management* 44(3), 411–27.

Shiva, V. (1993) *Monocultures of the Mind: perspectives on biodiversity and biotechnology*. Zed Books, London.

Shiva, V., Anderson, P., Schucking, H., Gray, A., Lohmann, L. and Cooper, D. (1991) *Biodiversity: social and ecological perspectives*. Zed, London.

Shrivastava, P. (1992) *Bhopal: anatomy of a crisis*. Paul Chapman, London.

Simon, J.L. (1981) *The Ultimate Resource* (2nd edn 1996). Princeton University Press, Princeton (NJ).

Simmons, I.G. (1989) *Changing the Face of the Earth: culture, environment, history*. Basil Blackwell, Oxford.

Simmons, I.G. (1990) Ingredients of a green geography. *Geography* 75(2), 98–105.

Simonis, U.E. (1990) *Beyond Growth: elements of sustainable development*. Editions Sigma, Bonn.

Simpson, E.S. (1987) *The Developing World: an introduction*. Longman, Harlow.

Sims, H. (1999) One-fifth of the sky: China's environmental stewardship. *World Development* 27(7), 1227–45.

Singh, R.B. (ed.) (2001) *Urban Sustainability in the Context of Global Change: towards promoting healthy and green cities*. Science Publishers Inc., Enfield (NH).

Smil, V. (1987) A perspective on global environmental crisis. *Futures* 19(1987), 240–53.

Smith. A. (1975) *An Enquiry into the Nature and Causes of the Wealth of Nations* (1st edn 1776). Dent, London.

Smith, K. (1992) *Environmental Hazards: assessing the risk and reducing disaster*. Routledge, London.

Smith, M. (1993) Cheney and the myth of postmodernism. *Environmental Ethics* 15(1), 3–18.

Smuts, J.C. (1926) *Holism and Evolution* (3rd edn 1936). Macmillan, London.

Sneddon, C. (2002) 'Sustainability' in ecological economics, ecology and livelihoods: a review. *Progress in Human Geography* 24(4), 521–49.

Socolow, R., Andrews, C., Berkhout, F. and Thomas, V. (eds) (1994) *Industrial Ecology and Global Change*. Cambridge University Press, Cambridge.

Soroos, M.S. (1997) *The Endangered Atmosphere: preserving the global commons*. University of South Carolina Press, Columbia (SC).

Sorrell, R.D. (1988) *St Francis of Assisi and Nature: tradition and innovation in western Christian attitudes towards the environment*. Oxford University Press, New York (NY).

Soussan, J.G. (1992) Sustainable development, in A. Mannion and S.R. Boulby (eds) *Environmental Issues in the 1990s.* Wiley, Chichester, pp. 21–36.

Stauffer, J. (1998) *The Water Crisis: constructing solutions to freshwater pollution.* Earthscan, London.

Stebbing, E.P. (1938) The advance of the Sahara. *The Geographical Journal* 91 356–9 .

Steffen, W., Sanderson, A., Tyson, P.D., Jäger, J., Matson, P.A., More, III, B., Oldfield, F., Richardson, K., Schellnhuber, H.-J., Turner, II, B.L. and Wasson, R.J. (2004) *Global Change and the Earth System: a planet under pressure.* Springer-Verlag, Berlin.

Stikker, A. (1998) Water today and tomorrow. *Futures* 30(1998), 43–62.

Stolton, S. and Dudley, N. (eds) (1999) *Partnerships for Protection: new strategies for planning and management of protected areas.* Earthscan, London.

Stone, R.D. (1992) *The Nature of Development: a report from the rural tropics on the quest for sustainable economic growth.* A.A. Knopf, New York (NY).

Stott, P. (2003) You can't control climate. *New Scientist* 177(2416), 25.

Stott, P. and Sullivan, S. (eds) (2000) *Political Ecology: science, myth and power.* Arnold, London.

Stren, R.E. and White, R. (eds) (1992) *Sustainable Cities: urbanization and the environment in international perspective.* Westview Press, Boulder (CO).

Stretton, H. (1976) *Capitalism, Socialism and the Environment.* Cambridge University Press, Cambridge.

Strohm, L.A. (2002) Pollution havens and the transfer of environmental risk. *Global Environmental Politics* 2(2), 29–36.

Stuart, R. (2000) Environmental management systems in the 21st century. *Chemical Health and Safety* November–December, 23–5.

Stuart, R. and Evans, D. (2002) Use of life cycle assessment in environmental management. *Environmental Management* 29(10), 132–42.

Sullivan, C. (2002) Calculating a water poverty index. *World Development* 30(7), 1195–210.

Suriyakumaran, C. (1979) Environmental management for development, in UNAPDI (ed.) *The Environmental Dimensions in Development* (Proceedings of the Consultative Meeting on Methodology and Techniques for Identification and Incorporation of Environmental Dimensions in Development Planning. UN Asian and Pacific Development Institute, Bangkok, pp. 53–64.

Swain, A. (2001) Water wars: fact or fiction? *Futures* 33(2001), 769–81.

Swanson, T.M. and Barbier, E. (eds) (1992) *Economics for the Wilds: wildlife, wildlands, diversity and development.* Earthscan, London.

Swingland, I. (2003) *Capturing Carbon and Conserving Biodiversity: the market approach.* Earthscan, London.

Sylvan, R. and Bennett, D. (1988) Taoism and deep ecology. *The Ecologist* 18(4/5), 148–58.

Tait, J. and Morris, D. (2000) Sustainable development of agricultural systems. *Futures* 32(2000), 247–60.

Tawney, R.H. (1954) *Religion and the Base of Capitalism* (1st edn 1926). Mentor Books, New York (NY).

Teilhard de Chardin, P. (1964) *The Future of Man.* Collins, London.

Teilhard de Chardin, P. (1965) *The Phenomenon of Man.* Harper and Row, New York (NY).

Thana, N.C. and Biswas, A.K. (ed.) (1990) *Environmentally Sound Water Management.* Oxford University Press, Delhi.

Thérivel, R. and Partidário, M.R. (1996) *The Practice of Strategic Environmental Assessment.* Earthscan, London.

Thérivel, R. Wilson, E., Thompson, S., Heany, D. and Pritchard, D. (1992) *Strategic Environmental Assessment.* Earthscan, London.

Thomas, A.R. (1996) What is development management? *Journal of International Development* 8(1), 95–110.

Thomas, D.S.G. and Middleton, N.J. (1994) *Desertification: exploding the myth*. Wiley, Chichester.

Thomas, K. (1983) *Man and the Natural World: changing attitudes in England 1500–1800*. Penguin, Harmondsworth.

Thomas, W.L. (ed.) (1956) *Man's Role in Changing the Face of the Earth*. University of Chicago Press, Chicago (IN).

Thomas-Hope, E.M. and Hodgkiss, A.G. (1983) *A Geography of the Third World*. Methuen, London.

Thompson, D. (ed.) (2002) *Tools for Environmental Management: a practical introduction and guide*. New Society Publishers, Gabriola Island (BC, Canada).

Thompson, D. (1994) Environmental auditing theory and practice. *Environmental Management*, 18(4), 605–15.

Thompson, M., Warburton, M. and Hatley, T. (1986) *Uncertainty on a Himalayan Scale*. Ethnographia Press, London.

Thoreau, H.D. (ed.) (1960) *Walden, or Life in the Woods* (1st edn 1854, Houghton Mifflin, Boston). New American Library, New York (NY).

Tisdell, C. (1999) *Biodiversity, Conservation and Sustainable Development: principles and practices with Asian examples*. Edward Elgar, Cheltenham.

Todaro, M.P. (1989) *Economic Development in the Third World* (4th edn). Longman, Harlow.

Tokar, B. (1988) Social ecology, deep ecology and the future of green political thought. *The Ecologist* 18(4/5), 132–42.

Toteng, E.N. (2001) Urban environmental management in Botswana: toward a theoretical explanation of public policy failure. *Environmental Management* 28(1), 19–30.

Toye, J. (1987) *Dilemmas of Development: reflections on the counter-revolution in development theory and policy*. Blackwell, Oxford.

Toynbee, A. (1972) The religious background to the present environmental crisis. *International Journal of Environmental Studies* 3, 141–6.

Tribe, J., Font, X., Griffiths, N., Vickery, R. and Yale, K. (2000) *Environmental Management for Rural Tourism and Recreation*. Cassel, London.

Tribe, X.I. (2000) *Forest Tourism and Recreation: case studies in environmental management*. CABI Publishing, Wallingford.

Tudge, C. (1977) *The Famine Business*. Penguin, Harmondsworth.

Turnbull, J. (2004) Exploring complexities of environmental management in developing countries: lessons from the Fiji Islands. *The Geographical Journal* 170(1), 64–77.

Turner, K. and Jones, T. (eds) (1991) *Wetlands: market intervention failures, four case studies*. Earthscan, London.

Turner, R.K. (ed.) (1988) *Sustainable Environmental Management: principles and practice*. Westview Press, Boulder (CO).

Turner, R.K. (ed.) (1993) *Sustainable Environmental Economics and Management: principles and practice*. Belhaven, London.

Turner, R.K., van den Bergh, J.C.L.M. and Bruwer, R. (eds) (2003) *Managing Wetlands: an ecological economics approach*. Edward Elgar, Cheltenham.

Uglow, J. (2002) *The Lunar Men: the friends who made the future*. Faber and Faber, London.

UNCHS (1988) *Refuse Collection Vehicles for Developing Countries*. UNCHS (Habitat), Nairobi (Kenya).

UNCTAD (1993) *Environmental Management in Transnational Corporations: report on the Benchmark Corporate Environmental Survey*. United Nations Conference on Trade and Development, United Nations, New York (NY).

UNDP (1991) *Human Development Report 1991*. Oxford University Press, Oxford.

UNEP (1994) *Convention on Biodiversity: Preamble* (UNEP/CBD/04/0). UN, Geneva.

Uphoff, N. (ed.) (2002) *Agroecological Innovations: increasing food production with participatory development*. Earthscan, London.

Utting, P. (ed.) (2004) *The Greening of Business in Developing Countries: rhetoric, reality and prospects*. Zed, London.

Van den Bergh, J. (1996) *Ecological Economics and Sustainable Development*. Edward Elgar, Cheltenham.

Van der Sluijs, J.P. (2002) Integrated assessment, in R.E. Munn (ed.) *Encyclopedia of Global Environmental Change*, vol. 4. Wiley, Chichester, pp. 250–3.

Vannucci, M. (ed.) (2001) *Mangrove Management and Conservation: present and future*. United Nations University Press, Tokyo.

Van Weizsacker, E.U. and Jessinghaus, J. (1992) *Ecological Tax Reform*. Zed, London.

Varley, A. (ed.) (1994) *Disasters, Development, and Environment*. Wiley, Chichester.

Velva, V., Hart, M., Greiner, T. and Crumbley, C. (2001) Indicators of sustainable production. *Journal of Cleaner Production* 9(5), 447–52.

Verheem, R.A.A. and Tonk, J.A.M.N. (2000) Strategic environmental assessment: one concept, multiple forms. *Impact Assessment and Project Appraisal* 18(3), 177–82.

Véron, R. (2001) The "New" Kerala Model: lessons for sustainable development. *World Development* 29(4), 601–17.

Victor, P.A. (1991) Indicators of sustainable development: some lessons from capital theory. *Ecological Economy* 4, 191–213.

Vogt, W. (1948) *Road to Survival*. William Sloane, New York (NY).

Von Moltke, K. (1996) *International Environmental Management, Trade Regimes and Sustainability*. International Institute for Sustainable Development, Winnipeg.

Wackernaegel, M. and Yount, J.D. (2000). Footprints for sustainable development: the next steps. *Environment, Development and Sustainability* 2(1), 21-42.

Wackernaegel, M. and Rees, W. (2003) *Our Ecological Footprint: reducing human impact on the Earth*. New Society Publishers, Gabriola Island (BC, Canada).

Waddington, C.H. (1977) *Tools for Thought*. Jonathan Cape, London.

Wakefield, J. (2002) Boys won't be boys. *New Scientist* 174(2349), 42–5.

Wall, D. (1994) *Green History: a reader in environmental literature, philosophy and politics*. Routledge, London.

Wallart, N. (1999) *The Political Economy of Environmental Taxes*. Edward Elgar, Cheltenham.

Walters, C. (1996) *Adaptive Management of Renewable Resources*. Macmillan, New York (NY).

Ward, B. (1976) *The Home of Man*. Penguin, Harmondsworth.

Ward, B. and Dubos, R.E. (1972) *Only One Earth: the care and maintenance of a small planet*. Penguin, Harmondsworth.

Ward, C. (ed.) (1974) *Peter Kropotkin: fields, factories and workshops tomorrow*. Allen and Unwin, London.

Ward, P.M. (1990) *Mexico City: the production and reproduction of an urban environment*. Belhaven, London.

Warford, J. and Partow, Z. (1989) Evolution of the World Bank's environmental policy. *Finance & Development* December, 5–9.

Warhurst, A. (ed.) (1999) *Mining and the Environment: case studies from the Americas*. International Development Research Centre, P.O. Box 8500, Ottawa, Canada.

Warren, A. (2002) Land degradation is contextual. *Land Degradation & Development* 13(1), 1–11.

Warwick, C.J., Mumford, J.D. and Norton, G.A. (1993) Environmental management expert systems. *Journal of Environmental Management* 39, 251–70.

Waterbury, J. (2002) *The Nile Basin: national determinants of collective action*. Yale University Press, Newhaven (CT).

WCD (2004) *Dams and Development: a new framework for decision-making* (Report of the World Commission on Dams). Earthscan, London (available online at http://www.dams.org).

Weaver, G. (1999) *Strategic Environmental Management: using TQEM and ISO14001 for competitive advantage*. Wiley, New York (NY).

Weber, M. (1958) *The Protestant Ethic and the Spirit of Capitalism* (trans. T. Parsons). Charles Scribner, New York (NY) (Original German edn 1904 and 1905 as 2 vols.).

Weir, D. (1988) *The Bhopal Syndrome: pesticides, environment and health*. Earthscan, London.

Welford, R. (ed.) (1996a) *Corporate Environmental Management: systems and strategies* (vol. 1) (2nd edn). Earthscan, London.

Welford, R. (1996b) Regional development and environmental management: new opportunities for co-operation. *Scandinavian Journal of Management* 12(3), 347–57.

Welford, R. (1997) *Hijacking Environmentalism*. Earthscan, London.

Welford, R. (2000) *Corporate Environmental Management: towards sustainable development* (vol. 3). Earthscan, London.

Wellburn, A. (1988) *Air Pollution and Acid rain: the global threat of acid pollution*. Earthscan, London.

White, G. (1977) *A Natural History of Selborne* (1st edn 1789). Penguin, Harmondsworth.

White, jnr, L. (1967) The historical roots of our environmental crisis. *Science* 155(3767), 1203–7.

White, R. and Whitney, J. (eds) (1992) *Sustainable Cities: urbanisation and the environment in international perspective*. Westview Press, Boulder (CO).

Whittow, J. (1980) *Disasters: the anatomy of environmental hazards*. Pelican Books, London.

Whyte, I.D. (1996) The impact of global warming: sea level rise, in I.D. Whyte (ed.) *Climate Change and Human Society*. Arnold, London, Chapter 5.

Whiteside, M. (1998) *Living Farms: encouraging sustainable smallholders in Southern Africa*. Earthscan, London.

Wijkman, A. and Timberlake, L. (1984) *Natural Disasters: acts of God or acts of man?* Earthscan, London.

Wilkie, D.S. and Carpenter, J. (1999) Can nature tourism help finance protected areas in the Congo Basin? *Oryx* 33(4), 333–9.

Wilkinson, G. and Dale, B.G. (1999) Integration of quality, environmental and health and safety management systems: an examination of the key issues. *Proceedings of the Institution of Mechanical Engineers* 213B, 275–83.

Wilson, E.O. (2002) *The Future of Life*. Little, Brown & Co., Boston (MD).

Wilson, G.A. and Bryant, R.L. (1997) *Environmental Management: new directions for the twenty-first century*. University College London Press, London.

Winder, C. (1997) Integrating Quality, Safety, and Environmental Management Systems. *Quality Assurance* (San Diego, CA), 5(Jan–Mar), 27–48.

Wood, A., Steadman-Edwards, P. and Mang, J. (eds) (2000) *The Root Causes of Biodiversity Loss*. Earthscan, London.

Wood, C. and Djeddour, M. (1992) Strategic environmental assessment: EA of policies, plans and programmes. *Impact Assessment Bulletin* 10(1), 3–22.

World Bank (2000) *The Little Green Data Book*. World Bank, Washington (DC).

World Bank (2002) *Environment Matters* July 2001–July 2002, 21.

World Bank (2003) *World Development Report 2003. Sustainable Development in a Dynamic World: transforming institutions, growth, and quality of life*. Oxford University Press, Oxford.

World Commission on Environment and Development (1987) *Our Common Future* (the Bruntland Report). Oxford University Press, Oxford.

World Resources Institute (1996) *World Directory of Country Environmental Studies: an annotated bibliography of natural resource profiles, plans and strategies* (produced by its International Environmental and Natural Resource Assessment Information Service). World Resources Institute, Washington (DC).

Wright, S.D. (2000) Grey or green? Stewardship and sustainability in an ageing society. *Journal of Ageing Studies* 14(3), 229–49.

Yang, H and Zehnder, A.J.B. (2002) Water scarcity and food import: a case study for southern Mediterranean countries. *World Development* 30(8), 1413–30.

Young, E. (2004) Taboos could save the seas. *New Scientist* 182(2443), 9.

Young, J. (1990) *Post Environmentalism*. Belhaven, London.

Young, Z. (2002) *A New Green Order?* Pluto Press, London.

Zimmerer, K.S. and Bassett, T.J. (eds) (2003) *Political Ecology: an integrative approach to geography and environment–development studies*. Guilford Publications, New York (NY).

Index

Entries are for acronyms, followed by the full name in brackets, e.g. AOSIS (Association of Small Island States). 'Box' or 'Fig.' indicate Boxes or Figures in the text, e.g. 87 Box 4.1